Principles of Light Vehicle Air Conditioning

Preface

Welcome to Principles of Light Vehicle Air Conditioning

This book has been written because, as the number of vehicles on the world's roads rises, the demand for increased levels of comfort and convenience also grows.

While air conditioning and climate control may be seen as a luxury by some, the key benefits often outweigh the initial costs and resources required to implement these systems on newly produced vehicles; in fact most new cars come with some form of air conditioning as standard.

An environment which helps keep the driver and passengers comfortable and alert, maintaining the correct levels of ventilation and humidity, can increase concentration and the ability to devote more of their attention to the occupation of driving.

The downside of these systems is the environmental impact of the chemicals used to provide the refrigeration process.

Globally, anthropogenic, or 'man-made' emissions are believed to be the key factor in climate change and refrigerants have a larger influence than many others.

Small amounts of fluorinated gasses released to atmosphere may be causing irreparable damage to our planet, initiating ozone depletion and global warming.

Although many organisations are currently seeking alternatives to these harmful cocktails, at the present time we are restricted by the availability, cost and technology required to make viable replacements.

This means that for the time being, technicians and air conditioning professionals need to ensure that refrigerants are handled with due diligence and systems are maintained to the highest standards in order to contain and reduce emissions. Remember these chemicals only become dangerous when released to atmosphere.

This book will help provide a knowledge and understanding of air conditioning and climate control, giving you the opportunity to work on these systems using prescribed methods and techniques proven to reduce the accidental release of refrigerants to atmosphere.

The chapters will introduce you to health and safety, refrigeration principles and the handling and diagnosis of air conditioning systems.

It also lays out key terms, points of interest, safety and diagnostic tips in order to support the information provided within the text.

Chapters:

Chapter 1 Refrigerant Handling.............**Page 3**

Chapter 2 Mobile Air Conditioning Principles..**Page 45**

Chapter 3 Diagnostics for Mobile Air Conditioning and Climate Control............................**Page 88**

This book offers:

Ideal support for learners and tutors undertaking automotive qualifications.

Information to help cover the knowledge requirements for refrigerant handling and climate control.

A large number of illustrations to support knowledge and understanding.

Text © Graham Stoakes 2015

Original illustrations © Graham Stoakes 2015

The rights of Graham Stoakes to be identified as author of this work have been asserted by them in accordance with the Copyright, Designs and Patents Act 1988.

Copyright notice ©

All rights reserved. No part of this publication may be reproduced In any form or by any means (including photocopying or storing it in any medium by electronic means and whether or not transiently or incidentally to some other use of this publication) without the written permission of the copyright owner, except in accordance with the provisions of the Copyright, Designs and Patents Act 1988 or under the terms of a licence issued by the Copyright Licensing Agency, Saffron House, 6 - 10 Kirby Street, London EC1N 8TS (www.cla.co.uk). Applications for the copyright owners' written permission should be addressed to the author.

Principles of Light Vehicle Air Conditioning

Acknowledgements

Graham Stoakes would like to thank Anita and Holly Stoakes for their support during this project.

Thank you to alerrandre for the cover design.

The author and publisher would also like to thank the following individuals and organisations for permission to reproduce photographs:

Shutterstock.com

Cover image: Shutterstock.com – Mr.Exen

Author

Graham Stoakes AAE MIMI QTLS is a lecturer and author of college textbooks in automotive engineering for light vehicles and motorcycles.

With his background as a qualified Master Technician, senior automotive manager and specialist diagnostic trainer, he brings over 30 years of technical industry experience to this title.

Cover design - fiver.com/alerrandre

Published by Graham Stoakes

First published 2015

First edition

ISBN 978-0-9929492-4-2

Refrigerant Handling

Chapter 1 Refrigerant Handling

This chapter will help you develop an understanding of the operation of air conditioning. It covers the principles behind these systems as well as an overview of their construction. This chapter also covers the environmental issues caused by refrigerants and describes methods that can be used to reduce their possible release to atmosphere. It supports you by providing knowledge that will help you when undertaking both theory and practical assessments. Remember to work safely at all times and observe the relevant environmental, health and safety regulations; while developing gas handling routines that are systematic and effective.

Contents

Basic operation of air conditioning systems in motor vehicles 6

Properties of refrigerants used in air conditioning systems in motor vehicles 8

Environmental impact of refrigerant gasses if released to atmosphere 14

Regulations and legislation 18

Procedures for recovering refrigerant gasses 31

Connection and operation of recovery sets 33

Safe working when handling refrigerant

There are many hazards associated with the handling of refrigerant used in air conditioning systems. You should always assess the risks involved with any maintenance or repair routine before you begin and put safety measures in place.

You need to give special consideration to the possibility of:
- Frostbite caused by the sudden release of refrigerant gas
- Suffocation caused by the venting of gas in a confined space

You should always use appropriate personal protective equipment (PPE) when you work on these systems. Make sure that your selection of PPE will protect you from these hazards.

Personal Protective Equipment (PPE)

Table 1.1 PPE required when working on vehicle air conditioning systems

PPE	Recommendations
Overalls	Overalls provide protection from coming into contact with oils and chemicals.
Gloves	Fluroelastomer gloves provide protection from fluorinated refrigerants and help protect the hands from frostbite.

Refrigerant Handling

Table 1.1 PPE required when working on vehicle air conditioning systems

PPE	Recommendations
Protective footwear 	Safety boots protect the feet from a crush injury and often have oil and chemical resistant soles. Safety boots should have a steel toe-cap and steel mid-sole.
Goggles 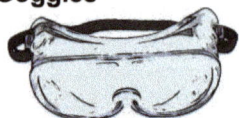	Safety goggles reduce the risk of small objects or refrigerants coming into contact with the eyes.
Bump cap/Hard hat 	A bump cap or hard hat protects the head from bump injuries when working under cars.

Vehicle Protective Equipment (VPE)

To reduce the possibility of damage to the car, always use the appropriate vehicle protection equipment (VPE):

Wing covers

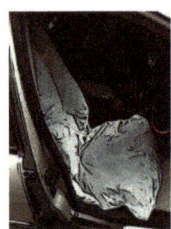
Seat covers

Steering wheel covers

Floor mats

Information sources

The complex nature of air conditioning and climate control systems requires you to have a good source of technical information and data. In order to conduct maintenance and repair procedures, you need to gather as much information as possible before you start.

Refrigerant Handling

Sources of information may include:

Table 1.2 Possible information sources

Verbal information from the driver	Vehicle identification numbers
Service and repair history	Warranty information
Vehicle handbook	Technical data manuals
Workshop manuals/Wiring diagrams	Safety recall sheets
Manufacturer specific information	Information bulletins
Technical helplines	Advice from other technicians/colleagues
Internet	Parts suppliers/catalogues
Jobcards	Diagnostic trouble codes
Oscilloscope waveforms	On vehicle warning labels/stickers
On vehicle displays	Temperature readings

Always compare the results of any inspection or testing to suitable sources of data. Remember that no matter which information or data source you use, it is important to evaluate how useful and reliable it will be to your safety, maintenance and repair routine.

Air conditioning and climate control

Air conditioning and climate control are comfort and convenience systems fitted to vehicles to help regulate:
Temperature
Ventilation
Humidity
Air purity
They are mainly based around refrigeration processes, but may be linked to a number of other systems and components to achieve all of the required features.

Air conditioning components

In order to function, the air conditioning system is made up of a number of components, table 1.3 illustrates main parts used and gives a brief description of their purpose.

Table 1.3 The main components of an air conditioning system

Air conditioning component and function	Example
Compressor - An engine or electrically driven pump, designed to raise the pressure of the system refrigerant.	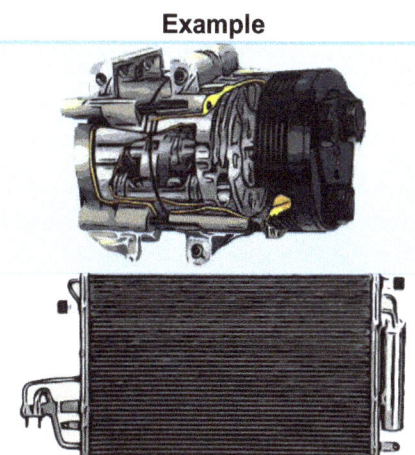
Condenser - A type of radiator, located outside of the vehicles passenger compartment (normally in front of the engine radiator). Its purpose is to cool hot refrigerant gas and allow it to condense into a liquid.	

Refrigerant Handling

Table 1.3 The main components of an air conditioning system

Air conditioning component and function	Example
Evaporator - A type of radiator, located inside the passenger compartment (normally just in front of the heater matrix). Its purpose is to allow the refrigerant to expand and evaporate, changing state from a liquid to a gas.	
Suction accumulator - A container designed to act as a temporary storage for liquid refrigerant. The suction accumulator allows the refrigerant to expand and evaporate before returning to the compressor. It will contain a silicone desiccant which is hygroscopic to help remove any water moisture present in the system. Receiver driers are used in fixed orifice tube FOT air conditioning systems.	
Fixed orifice tube FOT - This is an accurately sized restriction placed at the entry to the system evaporator. It ensures a constant metered flow of refrigerant can be maintained during air conditioning operation.	
Expansion valve - Also known as a thermal expansion valve or TXV, this is a variable sized nozzle which helps to control the amount of refrigerant which enters the evaporator. It will regulate the flow in accordance with evaporator temperature, allowing more refrigerant to enter as temperature rises and reducing refrigerant flow as temperature falls.	
Receiver drier - A container designed to act as a reservoir for liquid refrigerant. It will house a silicone desiccant which is hygroscopic to help remove any water moisture present in the system. Receiver driers are used in thermal expansion valve TXV air conditioning systems.	
Hoses - Rubber hoses help connect various parts of the system components together. Rubber is used to reduce the possibility of damage caused by vibrations between the components mounted solidly to the vehicle body and those mounted on the engine. To ensure correct operation they are made of rubber compounds which are resistant to chemical damage caused by refrigerants, and reduce the possibility of refrigerant leakage.	

Basic operating principles of air conditioning systems

Manufacturers use different types of air conditioning in their vehicle ranges, where refrigerant gasses are circulated in a closed circuit to help control passenger compartment temperature. Overall, their function and operation are very similar, and the two main designs are described in the next section:

Refrigerant Handling

Expansion valve type system (TXV)

The TXV air conditioning system uses an engine-driven or electrically operated pump called a compressor to raise the pressure of a refrigerant gas in a sealed system. The most common gas currently used is Tetrafluroethane, known as R134a.

From the compressor, the gas then passes through a radiator, called a condenser, which is normally mounted just in front of the cooling system radiator. Some of the heat caused by compression is removed and the high pressure gas is cooled slightly and condensed into a liquid. From here it is transferred into a storage container called a receiver drier until it is needed.

When the driver operates controls to lower the passenger compartment temperature of the car, the refrigerant is released through a temperature-controlled expansion valve (TXV). As the pressure falls, the liquid refrigerant changes state in another small radiator inside the car called the evaporator. The temperature in the evaporator falls, and as the cabin air is circulated through the fins of the evaporator, heat is removed. This helps cool the air inside the car. The refrigerant is then returned to the compressor, where the whole process starts once again.

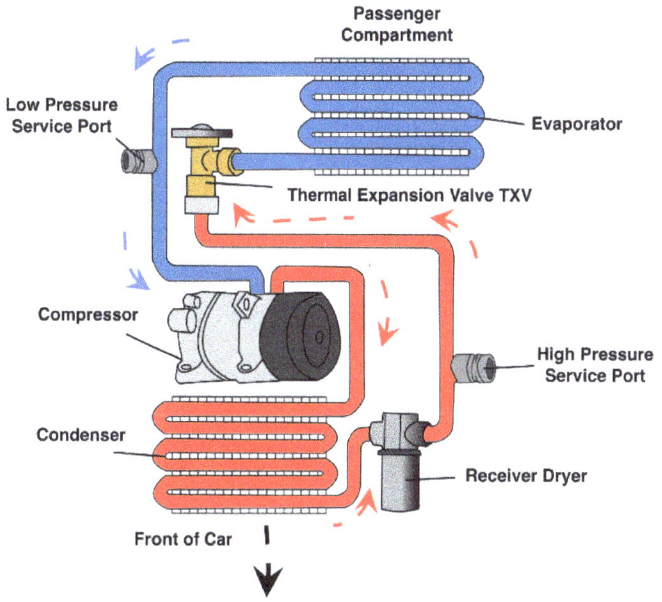

Figure 1.1 TXV air-conditioning circuit

Fixed orifice systems (FOT)

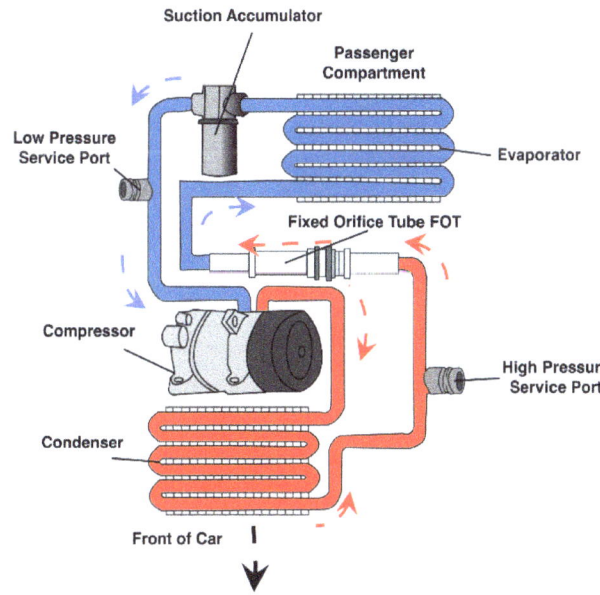

Figure 1.2 Fixed orifice air-conditioning circuit

An alternative air-conditioning system to the thermal expansion valve TXV type uses a process with a fixed orifice tube.

An engine-driven or electrically operated pump, called a compressor, is used to raise the pressure of a refrigerant gas in a sealed system. From the compressor, the gas then passes through a radiator, called a condenser, which is normally mounted just in front of the cooling system radiator. Some of the heat from compression is removed and the high pressure gas is cooled slightly and condensed into a liquid. After the condenser, the gas is passed through an accurately sized restriction called a 'fixed orifice' into the evaporator. In the evaporator, pressure falls allowing the refrigerant to change state back into a gas, and the temperature falls. As the cabin air is circulated through the evaporator fins, heat is removed. This helps cool the air inside the car.

When the cold gas leaves the evaporator, it enters a temporary reservoir called a suction accumulator, where any water moisture is removed and the refrigerant is stored before returning to the compressor, where the whole process starts all over again.

Refrigerant Handling

Key operating principles of air conditioning systems

Energy and heat

Heat is a form of energy and the scientific law of conservation states that 'energy cannot be created or destroyed, but it can be transferred or transformed from one form to another, including transformation into or from mass, as matter'. This means that the total amount of energy in a closed refrigeration system never changes.
Energy in a system can be transformed so that it changes into a different state, and energy in many states may be used to do a verity of physical work or perform functions; like cooling the passenger compartment of air conditioned vehicles.

An example of the transformation of energy can be shown when a vehicle is slowed down using brakes.
When a vehicle is in motion, it has kinetic energy. Kinetic energy is the energy of movement.
When the driver wants to slow down the kinetic energy of the vehicle needs to be removed. Because energy cannot be destroyed, it must be converted and transformed into another state, and this is done using the braking system.
As the driver applies the brakes, friction created by brake pads or brake shoes rubbing against the rotating surfaces at the wheels, transforms the kinetic energy into heat. This heat is then radiated to atmosphere, making the surrounding air a little warmer.
As there is now less kinetic energy in the vehicle, the car slows down.

Heat transfer

Because energy cannot be destroyed, if heat is added to something, it tries to spread out and fill up all the available space. Heat will always travel from the hotter to colder substance until an even quantity of heat energy exists, this is known as heat transfer.
Heat transfer takes place by three main methods:
Conduction - heat transfer through solids
Convection - heat transfer through a liquid or gas
Radiation - heat transfer through a vacuum or space

Radiation

Radiated heat is a type of electromagnetism. Electromagnetism takes many forms that are mainly invisible to the human eye. The different types of radiation can be found on a scale known as the 'electromagnetic spectrum'.
If it was visible, the electromagnetic spectrum would be seen as a waveform chart with the different types of radiation found at points with varying wavelength as shown in figure 1.3.

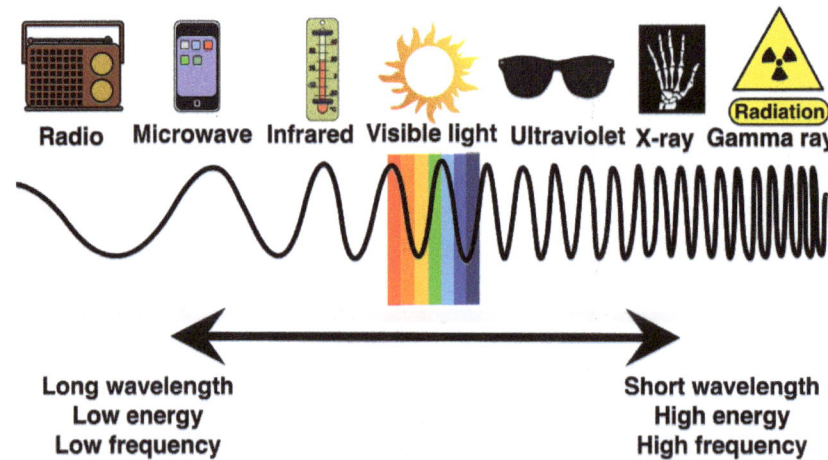

Figure 1.3 Radiation waveforms of the electromagnetic spectrum

Refrigerant Handling

Electromagnetic radiation will vary the amount and type of energy due to its wavelength; with radio waves having quite a low frequency and x-rays having quite a high frequency (see figure 1.3).
In the middle of the electromagnetic spectrum is visible light which can be further broken down into all the different colours of the rainbow from red to violet. On either end of the visible spectrum are inferred (IR) and ultra-violet (UV). Inferred is the main form of heat radiation, while ultra-violet can cause human cellular damage. Both these forms of radiation have a large environmental impact on planet earth and are affected by the use of air conditioning.

Humidity

Humidity is the relative amount of water moisture that is present in the air. It is often a product of warm air, but it is possible to have warm dry air which is known as 'arid'. The amount of moisture in the air is often called '**relative humidity**' and is given as a percentage compared to the maximum amount possible where the air becomes **saturated**. Humidity has a direct effect on comfort. The amount of moisture in the air can reduce the effectiveness of the body's ability to cool itself by sweating. Sweating works when perspiration on the surface of the skin can **evaporate,** and when humidity is high, evaporation cannot take place. Perspiration remains on the skin making you wet and clammy, but not really cooling. High humidity can also cause discomfort when breathing, especially in those with conditions such as asthma. Discomfort when breathing can lead to anxiety and loss of concentration.
Condensation is often a by-product of a high relative humidity environment. As the water moisture held in the warm air comes into contact with a cold surface, such as the inside of a car window, it will condense and cause the window to 'mist up'.

Figure 1.4 Condensation caused by humidity

You can often see the effects of humidity when having a cold drink. If the outside of the glass becomes wet, this is caused by water moisture from the air condensing on the cold surface.

Relative humidity - the amount of water moisture in the air compared to the maximum possible.

Saturated - when as much water as possible has been absorbed; thoroughly soaked.

Evaporate - to turn from a liquid to a vapour.

Condensation - water that collects as small droplets on a cold surface when humid air comes into contact with it.

Refrigerant Handling

Cold air is less able to hold on to moisture, and therefore is less humid. In fact, very cold countries will often have some of the driest air as water moisture content will be very low.

Air conditioning systems improve passenger comfort and concentration by reducing relative humidity. As cabin air passes through the air conditioning ventilation system, the water moisture will form **condensation** on the outer surface of the evaporator. The cooler air that is then returned to the passenger compartment is now far less humid. Operating the air conditioning can also speed up the de-misting of windscreens by operating it in combination with the vehicles heater system. Because the evaporator is mounted in front of the heater matrix, the cabin air is first cooled and dried, and then warmed up again. The result of this processes is warm dry air which helps clear windscreens; so air conditioning can be used all year round.

> Once the water moisture has condensed on the surface of the air conditioning systems evaporator, it is collected and drained from the bottom of the unit by small rubber or plastic pipes. These pipes lead to the underside of the car, and during operation, water will often drip out onto the ground. This is normal and can be an indication that the air conditioning system is functioning effectively.

Temperature

Temperature and heat are different.
Heat is the amount of energy in a substance and this energy is measured in Joules (J)
Temperature is a measurement of how hot or cold a substance is and is measured in degrees Centigrade (°C)
For example:

A kettle of water will be boiling at 100°C (this is the temperature), but will contain around 200KJ of energy (this is the heat).

A bath of water will be comfortable around 30°C (this is the temperature), but will contain around 50MJ of energy (this is the heat).

The sparks from a sparkler will each be around 1500°C (this is the temperature), but each spark will only contain around 2J of energy (this is the heat).

Figure 1.5 A boiling kettle at 100°C

Figure 1.6 A bath at 30°C

Figure 1.7 The sparks from a sparkler at 1500°C

This shows that it is the volume of a substance that has an effect on the amount of heat energy it can contain and not how hot or cold it is (i.e. its temperature).

Temperature is also often expressed in degrees Fahrenheit (°F) or Kelvin (K). Table 1.4 shows alternative values for the freezing and boiling points of water in Centigrade, Fahrenheit, and Kelvin.

Refrigerant Handling

Table 1.4 Temperature values for the freezing and boiling points of water

Water Freezes	Water Boils
Centigrade (°C) 0°C	100°C
Fahrenheit (°F) 32°F	212°F
Kelvin 273.15°K	373.15°K

Sensible and Latent heat

Sensible heat is the temperature range that a substance is able to remain as one state of matter, for example: Water is able to exist in the three main states of matter.

Solid - as ice at 0°C and below
Liquid - as water at 1° to 99°C
Gas - as steam at 100°C and above

At the point of change between solid to liquid and liquid to gas, an effect known as latent heat occurs and this is vital to the operation of air conditioning systems. To understand why, you need to know what happens at the point where a substance changes state.

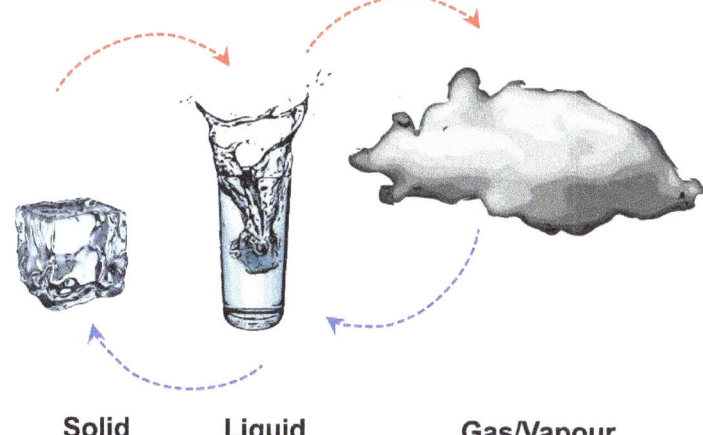

Solid   Liquid   Gas/Vapour

Figure 1.8 States of matter

The effects of Latent heat

Heating and ventilation engineers will often use British Thermal Units (BTU) to describe the heat energy in a substance, where:
One BTU is the amount of heat energy required to raise one pound of water (approximately a pint) by one degree Fahrenheit.
If water is used as an example of the effects of latent heat, then to raise the temperature of a pint of water by 1°F will take 1 BTU of thermal energy. To raise its temperature by another 1°F will take another 1 BTU and this is known as sensible heat.
When the pint of water reaches boiling point (which is 212°F) the amount of energy required to make it change state from a liquid to a gas rises to 970 BTU's. This large amount of extra heat energy has to come from somewhere, so it is absorbed from the surrounding air. This has the effect of removing heat energy from the air and making it feel colder.
As steam cools and condenses back into liquid water, heat energy is given up to the surrounding air warming it back up.
This change of state from liquid to gas and gas to liquid is vital in the operation of air to conditioning. As the refrigerant in the system turns from liquid to gas in the evaporator, it absorbs heat energy from the air circulated in the passenger compartment lowering the temperature. The refrigerant then continues on its cycle through the air conditioning system. When it reaches the condenser, on the outside of the car, heat is given up to atmosphere as the refrigerant turns from a gas to a liquid. In this way, the air conditioning system works as a heat exchanger, moving the heat energy from inside the car to the outside atmosphere.

Refrigerant Handling

Pressure

Pressure has a direct effect on the boiling point of a substance. As pressure is raised, boiling point increases and as pressure falls boiling point is reduced. This is very important in the operation of an air conditioning system as the refrigerants used have a very low boiling point. In order for the air conditioning system to be effective, refrigerants are used that are in a gaseous state at normal **ambient temperatures**, right down to 0°C and below. If during operation these refrigerants remained gas, there would be no change of state and no latent heat transfer. The gas would simply circulate within the system and no heat exchange would take place.

To ensure that the refrigerant is able to condense and become liquid, an engine driven or electrically operated compressor pump is used to raise system pressure. When the refrigerant is released into the evaporator, the rapid fall in pressure will instantly allow the refrigerant to boil and absorb the latent heat from the passenger compartment.

Ambient temperatures - the temperature of the immediate surroundings.

Pressure is force acting over the surface area of a substance and the pressure in an air conditioning system can be measured against the scale of atmospheric pressure or absolute pressure.
• Atmospheric pressure is a measurement of the natural air pressure occurring at sea level (1 bar, approximately 14.5 psi).
• Absolute pressure is a measurement of perfect vacuum and is often regarded as a negative value when compared to atmospheric pressure.

In order to test the operation of air conditioning, many system pressure gauges will show values of both positive (rising from 0 at atmospheric pressure) and negative (falling from 0 at atmospheric pressure down to a total vacuum).

The Système international (SI) unit of pressure is the Pascal (Pa) and is equal to one Newton of force per square meter. This compares with other units of pressure as shown in table 1.5.

Table 1.5 Units of pressure

Unit Abbreviation	Pascal Pa	Bar bar	Atmosphere atm	Pounds per square inch psi
1Pa	$1 N/m^2$	10^{-5}	9.8692×10^{-6}	145.04×10^{-6}
1bar	10^5	1.0bar	750.06	14.5037744
1atm	1.01325×10^5	1.01325	1.0atm	14.696
1psi	6.895×10^3	68.948×10^{-3}	68.046×10^{-3}	1.0psi

Pressures lower than atmospheric are often shown as a negative (-) value to represent vacuum. These can also be given a value in millimetres or inches of mercury (mm/Hg or in/Hg) which is defined as the pressure exerted by a column of mercury of 1 mm or inch in height at 0°C (32°F).

Refrigerant Handling

The refrigeration cycle

The refrigeration cycle of an air conditioning system is a closed loop, and as a result has no beginning or end. For the purpose of this description we will start and finish at the compressor.

Thermal expansion valve type (TXV)

1. The compressor pump is operated by the engine or an electric motor. Internal pistons or vanes are used to raise the pressure of the refrigerant gas (compressing the refrigerant will also raise its temperature). This system pressure increase will raise the boiling point of the refrigerant gas.
2. The refrigerant leaves the compressor as a hot high pressurised gas.
3. The refrigerant enters the top of the condenser radiator mounted on the outside of the vehicle. As the hot gas passes backwards and forwards across this radiator, air passing through the condenser will transfer some of the heat to the surrounding atmosphere. The fall in temperature of the refrigerant gas, as well as its high pressure, will allow the refrigerant gas to condense into a high pressure hot liquid (giving up a large amount of latent heat as it changes state).
4. From the condenser radiator, the hot, high pressure liquid refrigerant is transferred to a storage container known as a receiver drier. (Thermal expansion valve TXV system). The receiver drier also contains a silicone desiccant which will remove any water moisture.
5. From the receiver drier, a regulated amount of high pressure liquid refrigerant is allowed to pass through a valve into the evaporator unit, mounted inside the vehicles passenger compartment.
6. The evaporator is located in the low pressure/suction part of the air conditioning circuit. As the high pressure liquid refrigerant enters the low pressure space inside the evaporator, it instantly boils and turns into a low pressure gas (absorbing a large amount of latent heat as it changes state).
7. Passenger compartment air that is passed over the outside of the evaporator by an electrically driven fan has some of its heat energy removed and absorbed into the low pressure refrigerant gas.
8. The low pressure gas (with some heat energy from the passenger compartment) is then drawn back into the compressor where the refrigerant cycle begins all over again.

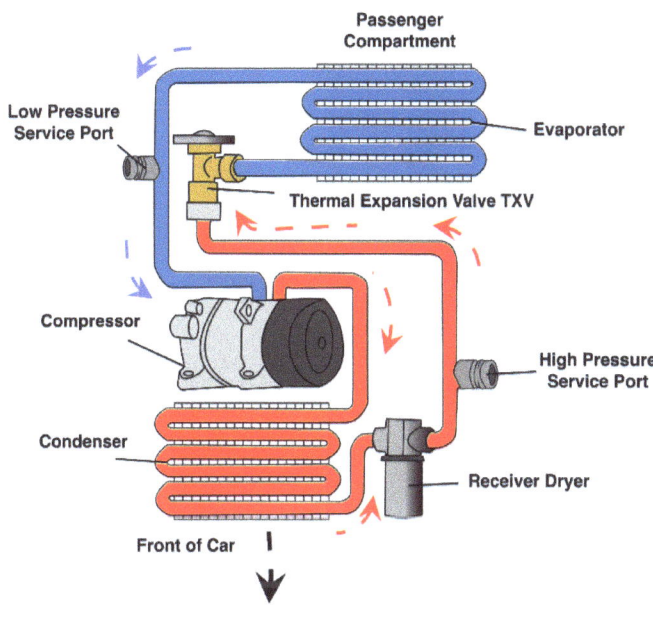

Figure 1.9 A TXV refrigeration circuit

A fixed orifice tube FOT follows the same refrigerant cycle, but instead of using a thermal expansion valve, it is passed through an accurately sized tube directly into the evaporator. To ensure that no refrigerant remains in a liquid state when it returns to the compressor, a storage container known as a suction accumulator collects excess liquid refrigerant, allowing it to fully evaporate before it continues its cycle. For a full description of the fixed orifice tube system see Chapter 2.

Refrigerant Handling

Alternative refrigeration cycle - R744 (Carbon dioxide)

1. The compressor pump is operated by the engine or an electric motor. Internal pistons or vanes are used to raise the pressure of the refrigerant gas to around 1,900 psi (compressing the refrigerant will also raise its temperature). This system pressure increase will raise the boiling point of the refrigerant gas.
2. The refrigerant leaves the compressor as a hot high pressurised gas.
3. The refrigerant enters the top of a gas cooler radiator mounted on the outside of the vehicle. As the hot gas passes backwards and forwards across this radiator, air passing through the gas cooler unit will transfer some of the heat to the surrounding atmosphere. The fall in temperature of the refrigerant gas is not enough to condense the carbon dioxide into a liquid at this point.
4. From the gas cooler radiator, the hot high pressure carbon dioxide passes through a heat exchanger pipe looping inside a cold suction accumulator. Some of the heat energy is released from the refrigerant and it now condenses into a high pressure liquid (see figure 1.10).
5. The high pressure liquid refrigerant is then transferred via pipes to a thermal expansion valve (TXV) or accurately sized restriction known as a fixed orifice tube (FOT). The flow of refrigerant to the FOT is controlled by regulating the system pressure, normally by switching the compressor on and off using a clutch mechanism.
6. From the fixed orifice tube, a regulated amount of high pressure liquid refrigerant is allowed to pass through into the evaporator unit, mounted inside the vehicles passenger compartment.
7. The evaporator is located in the low pressure/suction part of the air conditioning circuit. As the high pressure liquid refrigerant enters the low pressure space inside the evaporator, it instantly boils and turns into a low pressure gas (absorbing a large amount of latent heat as it changes state).
8. Passenger compartment air that is passed over the outside of the evaporator by an electrically driven fan has some of its heat energy removed and absorbed into the low pressure refrigerant gas.
9. After the refrigerant leaves the evaporator, it enters a storage container known as a suction accumulator. This contains a silicone desiccant which will remove any water moisture.
10. The low pressure gas (with some heat energy from the passenger compartment and some heat energy from the heat exchanger pipe) is then drawn back into the compressor where the refrigerant cycle begins all over again.

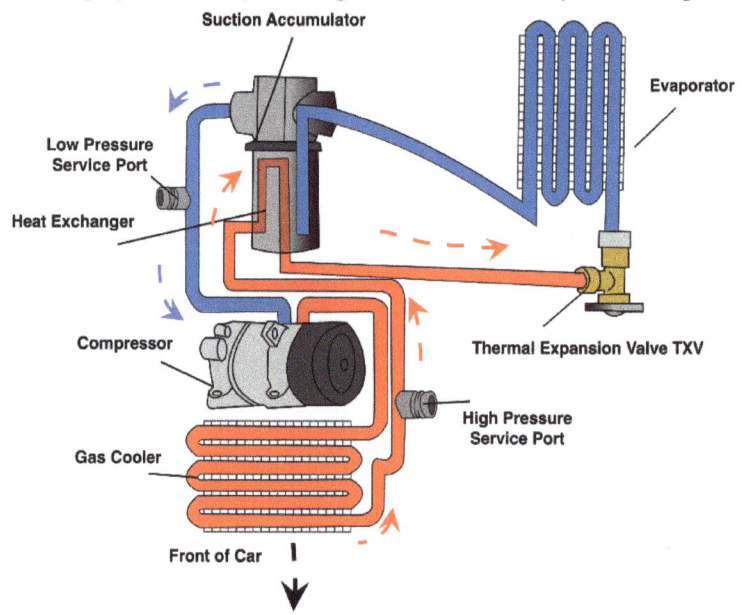

Figure 1.10 An R744 refrigeration circuit

Environmental issues relating to refrigerants

Early refrigeration systems used chemicals such as ammonia and sulphur dioxide. Both of these chemicals are known to cause health issues, particularly with breathing and lung function. Ammonia will also produce corrosive gasses which can damage human tissue.
To reduce the dangers posed by these substances, a chemist working for the Dayton Engineering Laboratory Corporation (DELCO) called Thomas Midgely Jr was tasked with finding an alternative refrigerant. He created a chemical compound known as Dichlorodifluoromethane (R12).

Refrigerant Handling

This refrigerant was initially thought to be totally harmless and safe. Unfortunately it was made from a blend of chlorofluorocarbons (CFC), and in the 1970's, with the help of satellite imagery, it was discovered that these chemicals were damaging the ozone layer. The amount of damage caused by a chemical to the ozone layer is rated as its ozone depletion potential (ODP).

Ozone layer and depletion

The ozone layer can be considered the planet's sunscreen, and without its protection, life on earth would not be possible. Ozone molecules block most of the harmful ultraviolet (UV) radiation from the sun. Ultraviolet radiation will still reach the surface of the earth but only in small quantities.

Small amounts of UV are beneficial for people and are essential in the production of vitamin D. Ultraviolet radiation is also used to treat several diseases, including rickets, psoriasis, eczema and jaundice.

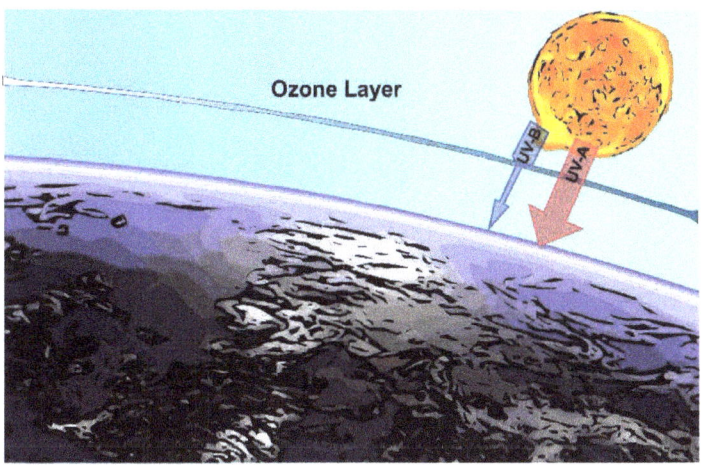

Figure 1.11 The ozone layer and ultraviolet radiation

Prolonged human exposure to solar UV radiation can result in acute and chronic health effects on the skin, eye and immune system. Sunburn is the best-known effect of excessive ultraviolet radiation exposure. Over the longer periods, UV radiation causes degenerative changes in cells of the skin, fibrous tissue and blood vessels leading to premature skin ageing and discolouration. Another long-term effect is an inflammatory reaction of the eye. In the most serious cases, skin cancer and cataracts can occur.

Ozone (O_3) is a gas that is generated naturally, mainly from the electric reaction of lightening with air. It rises up through the earth's atmosphere to the outer edge in an area known as the stratosphere. There it forms a very thin layer (around 2mm thick) which is enough to deflect most of the suns dangerous UV radiation.

The chemicals found in CFC refrigerants can also rise up into the stratosphere and damage the ozone layer. The sun's rays break the chemicals down into chlorine (Cl_2) and bromine (Br). One molecule of chlorine (Cl_2) can destroy 100,000 molecules of ozone (O_3), and therefore have a very high ozone depletion potential (ODP).

As the sun's rays contribute to the chemical reactions taking place between the chlorine and ozone, the depletion of the ozone layer tends to be worst at the planet's North and South poles, where long summer month's speed up the destruction. When viewed from space with satellite imagery, a hole in the ozone layer can clearly be seen. This hole used to come and go with the seasons, but as CFC use increased it has become permanent and will remain this way until the amount of these chemicals in our atmosphere are reduced.

Dichlorodifluoromethane (R12) has a lifespan of 102 years, meaning it will remain in the atmosphere for a considerable amount of time continuing to cause damage to the ozone layer.

Refrigerant Handling

In order to reduce the continued depletion of the ozone layer, the production and sale of CFC's has been heavily restricted worldwide. An alternative refrigerant gas had to be found to replace Dichlorodifluoromethane R12. Many air conditioning systems currently use Tetrafluoroethane R134a which is a hydro fluorocarbon HFC. This is still a fluorinated gas but the absence of chlorine means that it no longer has an ozone depletion potential ODP. It does however contribute to global warming.

Global warming and the greenhouse effect

Global warming is an environmental phenomenon which is causing the overall temperature of the planet to rise. Most scientists agree that global warming is happening, but often disagree by how much. Within the next 100 years, estimates range from between a 2°C and a 10°C increase.
It is thought that a rise of around 6°C could be enough to cause global extinction.
The main reason for global warming is the greenhouse effect.
The earth receives most of its energy in the form of solar radiation from the sun. A large proportion of this energy is in the form of visible light. Because of its wavelength, the electromagnetic radiation in the form of light is able to easily pass through the earth's atmosphere and when it reaches the surface, it warms the ground. At night, when it is dark, this heat energy is radiated back towards space as

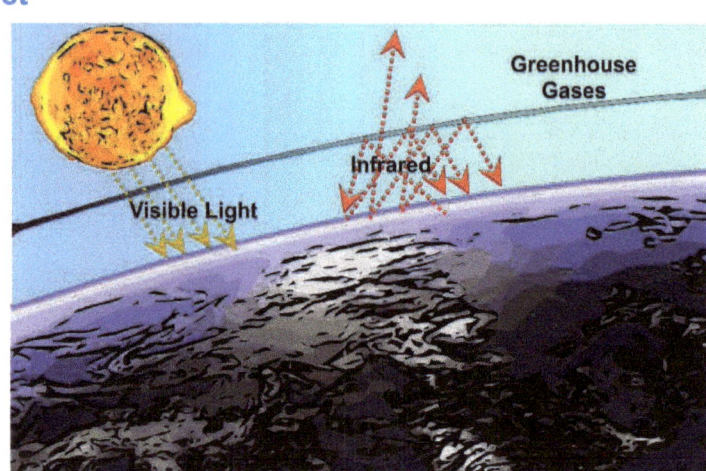

Figure 1.12 Global warming and the greenhouse effect

inferred, which has a slightly longer wavelength than visible light. Certain gasses in the earth's atmosphere are able to trap this inferred radiation adding to global warming. This process is known as the greenhouse effect.

The three main gasses which are known to contribute to the greenhouse effect are:

- Methane
- Water vapour
- Carbon dioxide (CO_2)

The largest known producers of methane on the planet are termites and cows.
The largest producers of carbon dioxide on the planet are humans, mainly through the burning of fossil fuels. (This is known as anthropogenic carbon dioxide).
Water vapour occurs naturally, and its effect on trapping heat radiation can be seen in night time temperatures. A clear night is cooler than a cloudy night, as more inferred radiation is able to escape into space.

Carbon dioxide is the base gas that all other types are compared to in order to judge their global warming potential. A molecule of carbon dioxide is given a global warming potential (GWP) of 1. In comparison to CO_2, the refrigerant gas Tetrafluoroethane (R134a) has a global warming potential of 1430. This means that one molecule of R134a is one thousand four hundred times more damaging to the environment than CO_2 and has a lifespan of around 13 years.

Refrigerant Handling

Table 1.6 shows the ozone depletion potential and global warming potential of possible refrigerants currently in use.

Table 1.6 Ozone depletion potential (ODP) and global warming potential (GWP) of refrigerants

Refrigerant	Ozone Depletion Potential (*ODP*)	Global Warming Potential (*GWP*)
R-11 Trichlorofluoromethane	1.0	4000
R-12 Dichlorodifluoromethane	1.0	2400
R-13 B1 Bromotrifluoromethane	10	-
R-22 Chlorodifluoromethane	0.05	1700
R-32 Difluoromethane	0	650
R-113 Trichlorotrifluoroethane	0.8	4800
R-114 Dichlorotetrafluoroethane	1.0	3.9
R-123 Dichlorotrifluoroethane	0.02	0.02
R-124 Chlorotetrafluoroethane	0.02	620
R-125 Pentafluoroethane	0	3400
R-134a Tetrafluoroethane	0	1430
R-143a Trifluoroethane	0	4300
R-152a Difluoroethane	0	120
R-245a Pentafluoropropane	0	-
R-401A (53% R-22, 34% R-124, 13% R-152a)	0.37	1100
R-401B (61% R-22, 28% R-124, 11% R-152a)	0.04	1200
R-402A (38% R-22, 60% R-125, 2% R-290)	0.02	2600
R-404A (44% R-125, 52% R-143a, R-134a)	0	3300
R-407A (20% R-32, 40% R-125, 40% R-134a)	0	2000
R-407C (23% R-32, 25% R-125, 52% R-134a)	0	1600
R-502 (48.8% R-22, 51.2% R-115)	0.283	4.1
R-507 (45% R-125, 55% R-143)	0	3300
R-717 Ammonia - NH_3	0	0
R-718 Water - H_2O	0	-
R-729 Air	0	-
R-744 Carbon Dioxide - CO_2	-	1
HFO-1234YF Tetrafluoropropene	0	<1

In order to combat ozone depletion and climate change, a number of treaties have been signed by some of the major countries around the world. They are designed to limit the production and use of certain chemicals which are known to produce environmental damage.

The two main treaties are:

The **Montreal Protocol** - focused around the area of ozone depletion
The **Kyoto agreement** - focused around the area of climate change

Refrigerant Handling

Montreal protocol

Montreal protocol is an international treaty designed to protect the ozone layer by phasing out the production of a number of substances believed to be responsible for ozone depletion. The treaty was opened for signature on September 16, 1987 and entered into force on January 1 1989. Following this, it has undergone seven revisions, in 1990 (London), 1991 (Nairobi), 1992 (Copenhagen), 1993 (Bangkok), 1995 (Vienna), 1997 (Montreal), and 1999 (Beijing). Since the Montreal protocol came into force, chlorinated emissions which damage the ozone layer have levelled off or even decreased in some cases. It is believed that if the international agreement is followed, the ozone layer may recover by 2050.

Kyoto agreement

Kyoto agreement is a protocol to the United Nations Framework Convention on Climate Change (UNFCCC). It is an international environmental treaty produced at the United Nations Conference on Environment and Development (UNCED), which was informally known as the Earth Summit.
The treaty is intended to achieve stabilisation of greenhouse gas concentrations in the atmosphere at a level that would prevent dangerous manmade (anthropogenic) interference with the climate system. The Protocol was initially taken up on 11 December 1997 in Kyoto, Japan, and entered into force on 16 February 2005. Under the Protocol, 37 countries have committed themselves to a reduction of four greenhouse gases - carbon dioxide, methane, nitrous oxide, sulphur hexafluoride and two further groups of gases - hydrofluorocarbons and perfluorocarbons (both of which may be used as refrigerants in air conditioning systems).

Legislation and regulations related to carrying out operations on air conditioning systems

Personnel carrying out certain operations (maintenance and repair) on mobile air conditioning (MAC) systems containing fluorinated (F) gas refrigerants must have appropriate qualifications. These operations include the removal of F gas refrigerants from all mobile equipment (excluding those in military use) when the air conditioning systems are under maintenance or prior to disposal of the equipment.
From 4th July 2010, all mobile air conditioning (MAC) technicians working with cars and car derived vans, must have achieved a minimum requirement of a refrigerant handling qualification which fulfils the European Union F Gas Regulation (EC842/2006 and Annex to Regulation EC307/2008).

Summary of the relevant provisions of regulation (EC) No 842/2006 and Directive 2006/40/EC

Regulation (EC) No 842/2006
The F gas Regulation was published on 14 June 2006 in the Official Journal of the European Union and entered into force on 4 July 2006. The key obligations applied with effect from 4 July 2007.
The objectives of the regulation are shown below, and an explanation of their meanings follows at the end of the section.

The principal objective is to contain, prevent and thereby reduce emissions of F gases covered by the Kyoto Protocol. This Regulation will make a significant contribution towards the European Community's Kyoto Protocol target by introducing cost-effective mitigation measures and to prevent distortion of the internal market.
The main focus is on containment and recovery of F gases, together with harmonised restrictions on the marketing and use of F gases in applications where containment of F gases is difficult to achieve or the use of F gases is considered inappropriate and suitable alternatives exist.

Refrigerant Handling

The sectors affected
The containment and recovery articles in the Regulation will have an impact on the commercial refrigeration, air conditioning and heat pump sectors and in the fire protection sector; and for the personnel involved in the installation, servicing and recovery of F gases from these systems as well as from equipment containing fluorinated greenhouse gas based solvents, high voltage switchgear and fire extinguishers. Operators of relevant systems will have a range of obligations including prompt leakage repair, leakage checking and record keeping and ensuring appropriately qualified personnel are used.

However, this Regulation will potentially also have an impact on a wider range of F gas uses due to the recovery obligation provided for in Article 4.3.

The Regulation will also impact on producers, importers and exporters of F gases if they produce, import or export more than 1 tonne of F gases per annum as they will have to report to the Commission and Member States' competent authorities on the amounts produced, imported or exported.

In addition, specified products and equipment that contain F gases will be subject to labelling requirements and specific uses of F gases and products that contain F gases are controlled or banned by the Regulation. These cover certain uses of sulphur hexafluoride for magnesium die-casing, use of certain F gases in non-refillable containers, fire protection systems, tyres, one component foams, novelty aerosols, footwear, windows and self-chilling cans.

Directive 2006/40/EC
During negotiations in Council it was agreed that the measures in the Regulation relating to Mobile Air Conditioning systems in motor vehicles (MAC) should form part of a separate Directive amending existing vehicle type approval legislation. The MAC Directive was published on 14 June 2006 in the Official Journal of the European Union and Member States shall adopt and publish by 4 January 2008 the laws, regulations and administrative provisions necessary to comply with this Directive. Member States shall apply those measures from 5 January 2008.

The MAC Directive sets out measures to minimise emissions of F gases from air conditioning systems in cars (or car derived vans). This is to be achieved principally through:
- The introduction of maximum leakage rates.
- The eventual phase out in MAC use of F gases with global warming potential greater than 150.

What this means:
- The regulation is designed to help European countries comply with their responsibilities under the Kyoto agreement without incurring excessive cost and effort.
- It is designed to control the storage and use of refrigerants until alternatives can be found.
- It affects people that work with and repair air conditioning systems on cars and car derived vans.
- It also affects companies that manufacture and store refrigerants. If they produce, store, import or export more than one tonne of refrigerant, they must register this with the appropriate local authorities.
- Labelling is required on equipment and storage containers to ensure it is correctly handled.
- Vehicle manufacturers have to ensure that any natural leakage from new air conditioning systems is adequately controlled. (Vehicles with one evaporator are not allowed to leak more than 40g of refrigerant in a one year period, and vehicles with two evaporators are not allowed to leak more than 60g refrigerant in a one year period).
- Alternative refrigerants with a global warming potential of less than 150 should be used where possible.

Environmental Protection Act 1990

The Environmental Protection Act 1990 (EPA) is an Act of Parliament in the United Kingdom that defines structure and authority for waste management and control of emissions into the environment. It also sets out a coordinated programme for regulating and licensing the acceptable disposal of controlled waste on land.

An enforcing authority can issue an enforcement notice or prohibition notice on someone who does not comply with the regulations and there are criminal penalties including fines and imprisonment for violations.

Section 33 of the Environmental Protection Act states that, no person may 'treat, keep or dispose of controlled waste in a manner likely to cause pollution of the environment or harm to human health'.

Refrigerant Handling

Section 34 of the Environmental Protection Act states that, 'any person who imports, produces, carries, keeps, treats or disposes of controlled waste or, as a broker, has control of such waste, to take all such measures applicable to him in that capacity as are reasonable in the circumstances'. This is known as a 'Duty of Care'

Duty of Care for waste

Waste is anything you own or produce, and you want, or are required to get rid of. The disposal of refrigerant will be classed as controlled waste and as a result, you, have a Duty of Care as to how this is done.
The Duty of Care is a law which says that you must take all reasonable steps to keep waste safe. If you give waste to someone else, you must be sure they are authorised to take it and can transport, recycle or dispose of it safely. If you break this law, you can be fined an unlimited amount.
The Duty of Care applies to anyone who produces or imports, keeps or stores, transports, treats or disposes of waste. It also applies if you act as a broker and arrange these things.

When you have waste you must:
- ✓ Store it safely and securely and ensure that it doesn't escape from your control.
- ✓ If you give waste to someone else, you must check they have authority to take it. The law says the person to whom you give your waste must be authorised to take it.
- ✓ You must describe the waste in writing. You must fill in and sign a transfer note for it. You must keep a copy of the transfer note.

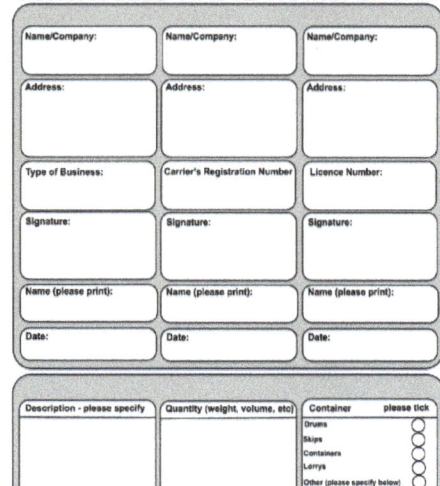

Figure 1.13 Waste transfer note

Both persons involved in the transfer of waste must keep copies of the transfer note and the description of the waste for two years.

Control of Substances Hazardous to Health Regulations 2002 (COSHH)

The handling and use of refrigerants will also come under the Control of Substances Hazardous to Health egulations COSHH 2002.

Some examples of hazardous substances are:
Natural/artificial
Liquid
Solid
Gas
Vapour
Dust

The degree of hazard for each substance should be indicated by a hazard symbol on the label.

Refrigerant Handling

Never use any hazardous substance unless you have received COSHH training. Always fully follow the procedures on how to use a hazardous substance.

Table 1.7 lists what you and your employer are responsible for under the Control of Substances Hazardous to Health Regulations 2002 (COSHH).

Table 1.7 Who's responsible for what under the Control of Substances Hazardous to Health Regulations 2002 (COSHH)

You are responsible for …	Your employer is responsible for …
Following the procedures put in place.	Completing risk assessments of all hazardous chemicals and substances used by employees.
Paying attention to training given.	Providing training for employees.
Wearing the correct PPE.	Putting steps in place to prevent exposure to hazardous chemicals and substances.

There are eight steps that employers must take to protect employees from hazardous substances. These are shown in the next section:

Step 1 • Find out what hazardous substances are used in the workplace and the risks these substances pose to people's health.

Step 2 • Decide what precautions are needed before any work starts with hazardous substances.

Step 3 • Prevent people being exposed to hazardous substances or, where this is not reasonably practicable, control the exposure.

Step 4 • Make sure control measures are used and maintained properly and that safety procedures are followed.

Step 5 • If required, monitor exposure of employees to hazardous substances.

Step 6 • Carry out health surveillance where assessment has shown that this is necessary or where COSHH makes specific requirements.

Step 7 • If required, prepare plans and procedures to deal with accidents, incidents and emergencies.

Step 8 • Make sure employees are properly informed, trained and supervised.

Figure 1.14 COSHH warning labels

Refrigerant Handling

A chemical is a single fluid, gas or solid in its natural state.
A substance is a mixture of chemicals.

Under the COSHH regulations, your organisation must carry out a risk assessment for every hazardous substance that is present in the workplace.
COSHH risk assessments must be carried out by a competent person with the legal and technical knowledge required. They will record the procedure that must be followed when using hazardous substances. If there are any changes to the procedure or different substances are introduced, then a new risk assessment should be carried out. This will make sure that the precautions for each hazard still sufficiently control the risk.

The risk assessment must include:
• The information on the supplier's **safety data sheet**
• How hazardous the substance is
• How much and how often it is used
• Who uses the substance

Safety data sheet – all the information required on a specific product to ensure it is used, stored and handled with care. The manufacturer must provide a safety data sheet with each product.

If any health problems occur or there are any defects in the control measures or PPE, you must report this to your employer immediately.

Types, use and properties of fluorinated gases

A number of gasses are in use as a refrigerant.
The main types that are often found in vehicle air conditioning systems are:
R134a - Tetrafluoroethane, this is currently the most common for use in vehicle air conditioning systems.
R12 - Dichlorodifluoromethane, this is now obsolete and vehicle technicians are no longer allowed to recover or use this refrigerant.
R22 - Chlorodifluoromethane, this is sometimes used as a substitute for R12 in older vehicles but requires special equipment and authorisation in order to be used.
R744 - Carbon dioxide, recommended for use in new in new air conditioning systems as it has a low GWP and does not contain fluorine.
R1234yf – Tetrafluoropropene, recommended for use in new in new air conditioning systems as it has a low GWP and does not contain carbon.

Refrigerant Handling

Notes: From October 2000, the sale of R12 was banned in the United Kingdom, and from 1st January 2001 vehicle technicians are no longer allowed to recover R12 from vehicles. If a system is suspected of containing R12, it must be recovered and recycled by an authorised specialist. The air conditioning system should then be converted (retrofitted) so that it is able to operate on R134a.

Whichever type of gas is used in a vehicle air conditioning system, it should be:
- Non-Flammable (although if R12 or R134a come into contact with a flame they can give off a highly toxic gas)
- Non-corrosive (although the chemicals in many refrigerants are able to damage certain types of rubber)
- Non-poisonous (although R134a if inhaled in large quantities, can replace the oxygen in the lungs leading to asphyxiation)
- Non-explosive (although any pressurised gas container has the potential for explosion if exposed to heat)
- Harmless to cloths and food (although it is not recommended to eat food that has come into contact with refrigerant gas)

Table 1.8 shows the different types of refrigerant commonly found in vehicle air conditioning systems and lists some of their properties.

Table 1.8 Refrigerants and their properties

Properties	Refrigerant				
	R12 (CFC)	R134a (HFC)	R22 (HCFC) sometimes used as an interim retrofit refrigerant between R12 & R134a	R744 (CO_2)	R-1234yf (HFO)
Chemical	Dichlorodifluoromethane	Tetrafluoroethane	Chlorodifluoromethane	Carbon dioxide	Tetrafluoropropene

Refrigerant Handling

Table 1.8 Refrigerants and their properties

Properties	Refrigerant				
	R12 (CFC)	R134a (HFC)	R22 (HCFC) sometimes used as an interim retrofit refrigerant between R12 & R134a	R744 (CO_2)	R-1234yf (HFO)
Hazards	Liquefied gas. In high concentrations may cause asphyxiation. Symptoms may include loss of mobility or consciousness. Victim may not be aware of asphyxiation. In low concentrations may cause narcotic effects. Symptoms may include dizziness, headache, nausea and loss of co-ordination.	In low concentrations may cause narcotic effects. Symptoms may include dizziness, headache, nausea and loss of co-ordination. In high concentrations may cause asphyxiation. Liquefied gas Not classified as dangerous substance.	Liquefied gas. In high concentrations may cause asphyxiation. May produce irregular heart beat and nervous symptoms. Gas/vapour heavier than air. May accumulate in confined spaces, particularly at or below ground level.	Liquefied gas. Low concentrations of CO2 cause increased respiration and headache. In high concentrations may cause asphyxiation. Symptoms may include loss of mobility or consciousness. Victim may not be aware of asphyxiation.	In low concentrations may cause narcotic effects. Symptoms may include dizziness, headache, nausea and loss of co-ordination. In high concentrations may cause asphyxiation. Symptoms may include loss of mobility or consciousness. Victim may not be aware of asphyxiation.
Fire hazards	Nonflammable. Exposure to fire may cause containers to rupture/explode. If involved in a fire the following toxic and/or corrosive fumes may be produced by thermal decomposition: Carbonyl fluoride. Carbon monoxide. Phosgene. Hydrogen chloride. Hydrogen fluoride.	Exposure to fire may cause containers to rupture/explode. Non flammable If involved in a fire the following toxic and/or corrosive fumes may be produced by thermal decomposition: Carbonyl fluoride.	Exposure to fire may cause containers to rupture/explode. Nonflammable. If involved in a fire the following toxic and/or corrosive fumes may be produced by thermal decomposition: Carbonyl fluoride. Carbon monoxide.	Exposure to fire may cause containers to rupture/explode. Nonflammable. All known extinguishers can be used.	Mild flammability (manageable). In case of fire hazardous decomposition products may be produced such as: Hydrogen fluoride. Carbonyl halides Carbon monoxide. Carbon dioxide (CO2).

Refrigerant Handling

Table 1.8 Refrigerants and their properties

Properties	Refrigerant				
	R12 (CFC)	R134a (HFC)	R22 (HCFC) sometimes used as an interim retrofit refrigerant between R12 & R134a	R744 (CO_2)	R-1234yf (HFO)
Environmental information	May have damaging effect on ozone layer. Covered by the 'Montreal Protocol'. ODP 1 GWP 10600	When discharged in large quantities may contribute to the greenhouse effect. ODP 0 GWP 1430	Covered by the 'Montreal Protocol'. May have damaging effect on ozone layer. When discharged in large quantities may contribute to the greenhouse effect. ODP 0.055 GWP 1700	When discharged in large quantities may contribute to the greenhouse effect. Can cause frost damage to vegetation. ODP 0 GWP 1	Global Warming Potential. When discharged in large quantities may contribute to the greenhouse effect. ODP 0 GWP 4 or <1 (depending on source information)
Boiling point	-29.8 °C	-26.3 °C	-40.9 °C	-78.5 °C	-29.4 °C

Many refrigerants are made up of a number of different chemicals (blends). Thermal decomposition is where the chemical substance is broken down into other chemicals through coming into contact with a source of heat. This is also known as fractionation.

Almost any condensable gas with a low boiling point will work as a refrigerant. It has been known for untrained people to use a hydrocarbon gas such as propane or butane in air conditioning systems. Although these gases will operate the refrigerant cycle adequately, they are also highly flammable and could cause an explosion in the event of an accident. To ensure that hydrocarbon contamination has not taken place, it is advisable to use a refrigerant identifier.

Health risks associated with air conditioning systems

There are many health and safety risks involved with working on an air conditioning system. These include:

Frostbite - if refrigerant is accidentally released from a pressurised cylinder or container, the sudden drop in pressure will cause it to boil and the change of state from liquid to gas will absorb latent heat. Any exposed body parts (skin and eyes for example) can instantly freeze causing frostbite.

Refrigerant Handling

Asphyxiation - many gasses used as refrigerants are not poisonous but if inhaled can displace the oxygen in the lungs. If released into a confined area, this could cause suffocation.

Thermal decomposition - although the gasses used for refrigerants are designed to be non-flammable, if they come into contact with an exposed flame or are drawn through the combustion process of an engine, their chemical consistency can be changed. The result is often highly toxic and corrosive chemicals that can affect the skin, eyes and lungs.

Incorrect use of hydrocarbon gasses - if the air conditioning system has been contaminated with a hydrocarbon gas, such as butane or propane there is an elevated risk of fire or explosion.

High system pressures - if the air conditioning system has become blocked or was over pressurised during repairs, a change in temperature could raise pressures to a point that could cause an explosion. For example, systems should be drained if the car is to be placed in a spray-bake oven following bodywork repairs.

Incorrect storage and use of refrigerant containers - refrigerants are stored safely as a high pressure liquid. The refrigerant containers should not be exposed to extremes of heat or cold as this could lead to rupture and explosion. They should be stored upright to ensure that no damage occurs to the valve mechanism, and they should not be stacked unless this is allowed by the container manufacturer.

Getting caught up in moving components/engine - as with any vehicle repairs, it will be necessary to test the operation of an air conditioning system. This will often involve the running of the engine and the dangers involved in working on and around rotating components.

Remember, if you are running an engine in a workshop to test the air conditioning system, you must use exhaust extraction.

Frostbite - Injury to body tissues caused by exposure to extreme cold. Skin tissue may not recover as it has been damaged by a form of 'cold burn'.

Asphyxiation – suffocation.

Table 1.9 Do's and Don'ts when working on air conditioning systems

Don't	Do
Weld, burn or carry out other hot work near air conditioning. The high temperatures may cause a pressure rise leading to an explosion.	Check vehicle manufacturer recommendations before the vehicle is placed in a spray-bake oven for painting. Air conditioning systems may need degassing before this is done.
Allow smoking, welding, burning or other hot work near the refrigerant, thermal decomposition will produce harmful gases.	Train staff in emergency actions in case of an accident with spilled refrigerant or gas leak.
Carry out roadside work on vehicles after an accident until the air conditioning system has been checked for damage/leaks and refrigerant type (hydrocarbons).	Check that all the refrigerant is removed from the air conditioning system before scraping the vehicle. This is known as end of life vehicle (ELV).

Refrigerant Handling

Table 1.9 Do's and Don'ts when working on air conditioning systems

Don't	Do
Overfill refrigerant containers (see how to calculate filling quantity).	Make arrangements for safe recovery and disposal of old or waste refrigerant. Section 33 & 34 Environmental Protection Act.
Mix R12 and R134a when recharging the system. Refrigerants are not interchangeable.	Always follow recommendations of vehicle and system manufacturers. Ensure adequate supervision and training before working on a system.
Work on air conditioning without noting risks and precautions.	Use approved equipment for maintenance, including recovery/recharge and leak detection.
Assume that the system is free from refrigerant gasses until proven. A zero pressure reading on the gauge just means no pressure, not no gas.	Identify what type of refrigerant is being used on the system before carrying out any work.
Allow the discharge of refrigerants into the atmosphere. Refrigerants can cause ozone depletion and global warming.	Wear appropriate eye protection, gloves and PPE when working on the system. Frostbite may cause permanent tissue damage.
Carry out repairs over or close to an inspection pit or work in a confined space. Refrigerant gas is heavier than air, and it can accumulate and cause asphyxiation.	Store refrigerant containers in a safe place away from direct heat and protect from frost.
Work on the compressor of a hybrid or electric vehicle without first isolating the high voltage system.	Use the correct non-electrically conductive compressor oil recommended by the manufacturer of any high voltage hybrid or electric vehicle system.

Personal protective equipment PPE

Having assessed any risks involved with the maintenance and repair of air conditioning systems, your health and safety can be further improved through the use of personal protective equipment.
Personal protective equipment (PPE) is the name for clothes and other items that you wear, which help provide defence against accidents or injury while you are carrying out your work. PPE is not the only way of preventing accidents or injury. It is your responsibility to make sure you are wearing your PPE when at work.
The use of PPE is covered by the Personal Protective Equipment (PPE) at Work Regulations 1992. This regulation requires that employers provide appropriate personal protective clothing and equipment for their employees.

Selecting the right PPE for the job

When selecting PPE, make sure that the equipment:
• is the right PPE for the job – ask for advice if you are not sure.
• fits correctly – it needs to be adjustable so it fits you properly.
• is properly looked after.
• prevents or controls the risk for the job you are doing.
• does not interfere with the job you are doing.
• does not create a new risk, e.g. overheating.
• is comfortable enough to wear for the length of time you need it.
• does not impair your sight, communication or movement.
• is compatible with other PPE worn – e.g. gloves and overalls do not leave exposed skin at the wrists where refrigerant might cause frostbite.

Refrigerant Handling

The following table shows the most common PPE recommended for use when working on and maintaining air conditioning systems.

Table 1.10 PPE and its uses

PPE	Use
Steel toe-capped boots 	Should always be worn in the workshop, in order to protect your feet. They have a reinforced toe area designed to protect against crush injury and will sometimes have a steel mid-sole to protect from punctures below. Most protective footwear will have an oil and chemical resistant sole.
Overalls 	Should always be worn in the workshop. Designed to keep oil, chemicals and dirt off of clothing and skin, it will also help to contain any loose clothing that may get caught up in moving engine components. If worn correctly, overalls will help to cover bare skin and reduce the possibility of frostbite if an accidental discharge of refrigerant occurs. If you are working on a high voltage electric or hybrid vehicle's air conditioning system, it is recommended that overall fastenings should be made of non-electrically conductive materials.
Nitrile or preferably Fluoroelastomer gloves 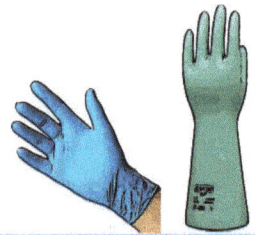	Gloves help protect your hands against the chemicals you may come Into contact with in an automotive workshop. In addition to latex and nitrile gloves, fluoroelastomer gloves are specifically designed to offer protection against the fluorinated gasses used as refrigerants. They should be worn in conjunction with overalls so that no bare skin is exposed at the wrists.
Goggles/Face shield  	Eye protection is vital when working on air conditioning and climate control systems. It is recommended that the protection should fully enclose the eyes to prevent contact in the event of an accidental discharge of refrigerant gas. Therefore goggles or a full face shield should be used in preference over safety glasses.

Refrigerant Handling

If you do not use the correct PPE you are potentially putting yourself at risk of severe and long-term health problems.

Checking and maintaining PPE and regulations

It is essential that your PPE is kept in good condition. If PPE is damaged, it will provide a lot less protection when you are wearing it.

Keep PPE in good condition by:
• Cleaning – for example, wash your overalls weekly or more often if working in dirty conditions.
• Examining – check for damage to equipment. For example, check that goggles fit well and that the lenses are clear.
• Replacing – if PPE becomes damaged, it must be replaced.

You can carry out simple maintenance of your PPE (for example, cleaning), but complicated repairs must only be done by an experienced person. It is your employer's responsibility to cover the cost of maintaining your PPE. Your employer must provide suitable storage for your PPE when it is not being used, unless you take the PPE away from the workplace (e.g. footwear or clothing). So it's your responsibility to use this storage space properly.

The CE mark found on PPE confirms that it has met the safety requirements of the Personal Protective Equipment at Work Regulations 1992. All PPE should have this mark.

Figure 1.15 The CE mark

Your employer cannot ask you for money for your PPE. Your responsibility as an employee is to look after the PPE and use it when required.
If you leave your place of work and you keep your PPE without your employer's permission, they (depending on your contract) might be able to deduct the cost of replacement PPE from your last pay.

Basic first aid

An automotive workshop is a high risk environment, and no matter what precautions are taken, there is always the possibility of accidents occurring which may lead to personal injury. The following advice is not a substitute for first aid training, and will only give you an overview of the action you may need to take. You should take care when you attempt to administer first aid that you do not place yourself in danger. Be very careful about what you do, because the wrong action can cause more harm to the casualty.

Good first aid always involves summoning appropriate help; many companies will have a trained first aider on site and must have a suitably stocked first aid box.

Refrigerant Handling

First aid box

The minimum level of first aid equipment in a suitably stocked first aid box should include:

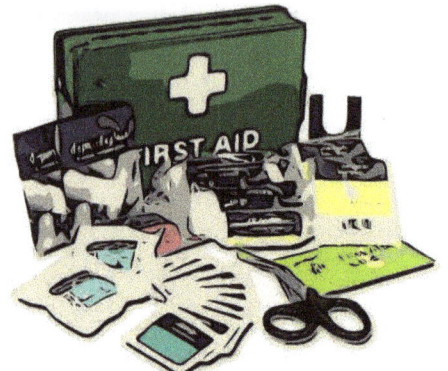

- A guidance leaflet.
- 2 sterile eye pads.
- 6 triangular bandages.
- 6 safety pins.
- 3 extra-large, 2 large and 6 medium-sized sterile unmedicated wound dressings.
- 20 sterile adhesive dressings (assorted sizes).
- 1 pair of disposable gloves (as required under HSE guidance).

Figure 1.16 First aid box

It is important to ensure that the contents of the first aid box are in date and are sufficient, based on the assessment of the workplace's first aid needs.

The law does not state how often the contents of a first aid box should be replaced, but most items, in particular sterile ones, are marked with expiry dates.

Other equipment such as eye wash stations must also be available if the work being carried out requires it.

Getting help

If you need to call for assistance, the main emergency services can be contacted by calling 999 free of charge from any landline or mobile phone.

When calling the emergency services, make sure you give the following information:

- Your telephone number.
- The location of the incident.
- The type of incident.
- The gender and age of the casualty.
- Details of any injuries observed.
- Any information you have observed about hazards, for example high voltage systems, chemical spills, gas or fuel leaks.

First aid measures for use with refrigerant

In the event of an accident occurring with refrigerant gasses, the following first aid measures may be appropriate:
Inhalation - Remove victim to uncontaminated area wearing self-contained breathing apparatus. Keep victim warm and rested. Call a doctor. Apply artificial respiration if breathing stopped.
Skin/eye contact - Immediately flush eyes thoroughly with water for at least 15 minutes. In case of frostbite spray with water for at least 15 minutes. Apply a sterile dressing. Obtain medical assistance.
Ingestion - Ingestion is not considered a potential route of exposure.

Refrigerant Handling

The recovery position

When dealing with health emergencies, you may need to place someone in the recovery position. In this position a casualty has the best chance of keeping a clear airway, not inhaling vomit and remaining as safe as possible until help arrives. You should not attempt to put someone in the recovery position if you think they might have back or neck injuries, and it may not be possible if any limbs are fractured.

Putting a casualty in the recovery position

1. Kneel at one side of the casualty, at about waist level.

2. Tilt the head back – this opens the airway. With the casualty on their back, make sure that their limbs are straight.

3. Bend the casualty's near arm so that it is at right angles to the body. Pull the arm on the far side over the chest and place the back of the hand against the opposite cheek.

4. Use your other hand to roll the casualty towards you by pulling gently on the far leg, just above the knee. This will bring the casualty onto their side.

5. Once the casualty is rolled over, bend the leg at right angles to the body. Make sure the head is tilted well back to keep the airway open.

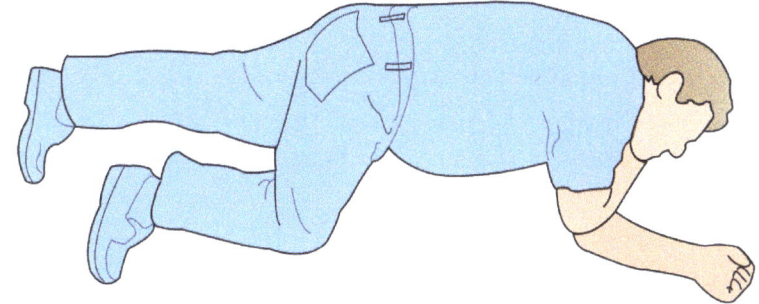

Figure 1.17 The recovery position

To find out more about first aid at work, visit the first aid section of the HSE website –
http://www.hse.gov.uk/firstaid/index.htm

Cylinder handing and recovery procedures

Refrigerant identification

It is very important to identify the type of refrigerant contained in an air conditioning system before any work is conducted or a recovery is started. Incorrect identification may lead to you braking the law, compromising safety and damaging equipment.

Various methods of identification include: Vehicle stickers and labels - normally found in the engine compartment, these stickers show the type of refrigerant used as well as the correct filling weights and system lubrication oil type.

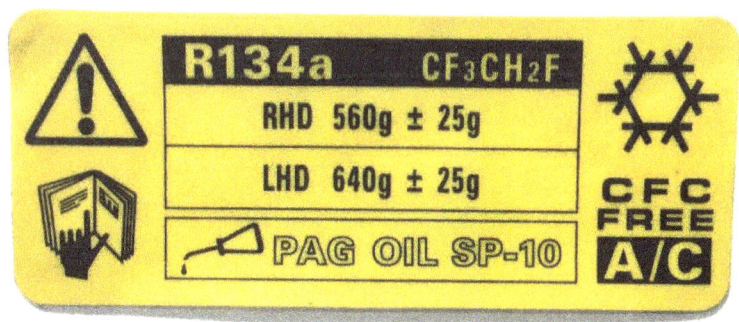

Figure 1.18 Vehicle stickers

Refrigerant Handling

Technical data manuals - these data manuals will show the recommended refrigerant type and filling weights. They will also often list the correct system lubrication oil type.

Service connectors - systems containing R134a will be fitted with quick fit/release service connections. These are different sizes to help avoid incorrect connection (high pressure has a large 16 mm service connection, low pressure has a small 13mm service connection). The service ports of a system containing R1234yf are similar to those found on R134a, however, they are slightly smaller in size to prevent misuse. Systems containing R12 are designed to have a screw type service connection which will not fit the recovery equipment used for R134a. The high pressure connection is a 3/16 of an inch flare and 3/8 - 24 thread type. The low pressure connection is a 1/4 of an inch flare and 7/16 - 24 thread type.

Figure 1.19 Service connectors

In theory it should not be possible to connect a recovery machine to the incorrect system type, due to the difference in style of the service connectors. Many machines are supplied with screw thread adapters for connections to a refrigerant cylinder. Without correct knowledge and training, these adapters could be used to incorrectly connect an R134a recovery machine to a system containing R12. If this were to happen, the operator would be breaking the law concerning the recovery of R12 and would also contaminate and damage the recovery equipment.

Hoses - the rubber hoses used to connect the various parts of the air conditioning system components will often have refrigerant type indicated on the outside. This is because certain rubbers will be damaged if the incorrect refrigerant type is used.

Refrigerant identifier - this is a small device of similar size to a multimeter. It has a hose and service connector to enable it to be connected to the air conditioning system. When operated, a small amount of refrigerant is drawn into the device, analysed and the type is then shown on a liquid crystal display screen.

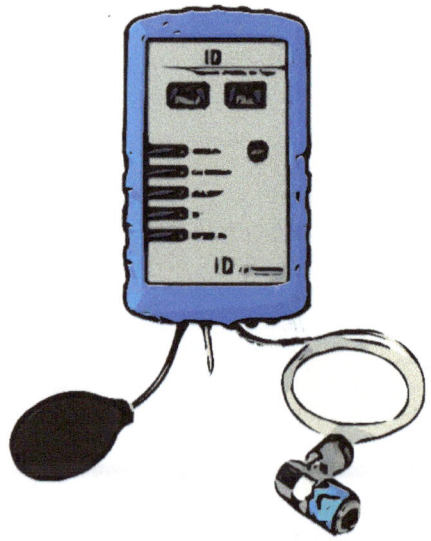

Figure 1.20 Refrigerant identifier

Refrigerant Handling

To help you identify the type of refrigerant stored in a cylinder, they are labelled and often colour coded. Table 1.11 shows the recommended colour coding for refrigerant cylinders and the contents.

Table 1.11 Refrigerant cylinder colour codes

Refrigerant code	Chemical Name	Cylinder Colour Code
R11	Trichlorfluoromethane	Orange
R12	Dichlorodifluoromethane	White
R13	Chlorotrifluoromethane	Light Blue
R113	Trichlorotrifluoroethane	Dark Purple
R114	Dichlorotetrafluoroethane	Navy Blue
R12/114	Dichlorodifluoromethane, Dichlorotetrafluoroethane	Light Grey
R13B1	Bromotrifluoromethane	Pinkish-Red
R-22	Chlorodifluoromethane	Light Green
R-23	Trifluoromethane	Light Blue Grey
R123	Dichlorotrifluoroethane	Light Blue Grey
R124	Chlorotetrafluoroethane	DOT Green
R134a	Tetrafluoroethane	Light Blue
R1234yf	Tetrafluoropropene	White with a Red Band

Note: it is worthwhile remembering that although these are the recommended cylinder colour codes, it may not be mandatory and some manufacturers rely on cylinder labelling.

Refrigerant cylinders will have a metal section around the main filling valve. This is to act as a protective shield for the valve to help prevent accidental damage. It also acts as a stand so that the cylinder can be inverted during refrigerant transfer procedures.

Refrigerant cylinders must:
- Be labelled to describe the contents
- Protected from sources of heat or frost
- Transported upright
- Not stacked on top of each other (unless allowed by manufacturer)

Service recovery units

Although the service and recovery equipment used to maintain an air conditioning system can be purchased and used as individual tools, many mobile air conditioning repairs are now conducted using fully automated machines known as recovery management stations (RMS). This has the advantage of the technician being able to connect and set up the equipment, and once in operation, they can do something else while recovery/service procedure is automatically conducted by the machine. The component parts of a service recovery machine are shown in table 1.12.

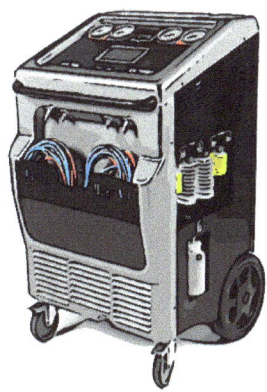

Figure 1.21 A recovery management station RMS

Refrigerant Handling

Table 1.12 The component parts of an air condition service/recovery machine.

Recovery machine component	Description
Hoses 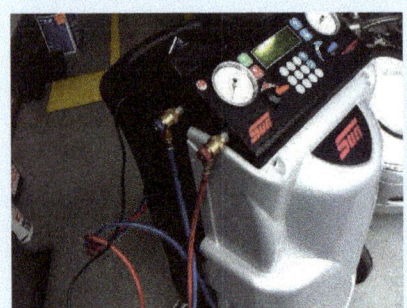	Two hoses are attached to the recovery machine, one red and one blue. These are for attachment to the high pressure (red) and low pressure (blue) service port connections on the vehicle. On the outer end they are equipped with quick fit connections which reduce the possibility of incorrect connection.
Manifold gauges and valves 	Two colour coded gauges are mounted on the front of the recovery machine, one blue (low pressure) and one red (high pressure). Each gauge will have a different range scale to show system pressures and will also have a reading below zero which represents a vacuum. Two colour coded taps will be included in the manifold gauge design which are able to isolate the air conditioning machine from the hoses and vehicle air conditioning system.
Storage tank 	The recovery machine will have its own internal storage tank for refrigerant gas. This is a working chamber, receiving refrigerant during a de-gas and supplying refrigerant during a re-gas. This storage tank should normally be kept around half full to allow space for system recovery and have an adequate quantity for system charging.
Heater belt 	Temperatures will have an effect on the speed and accuracy with which refrigerant gas can be transferred between the recovery machine and the air conditioning system. Some fully automatic recovery machines have an electrically operated heater belt which is wrapped around the storage container. As gas is transferred between the recovery machine and air conditioning system, the pressure change in the refrigerant will often lead to the absorption of latent heat and the storage container temperature can fall. This is measured by a sensor and the heater belt is switched on to maintain a relatively stable temperature.

Refrigerant Handling

Table 1.12 The component parts of an air condition service/recovery machine.

Recovery machine component	Description
Weighing scales 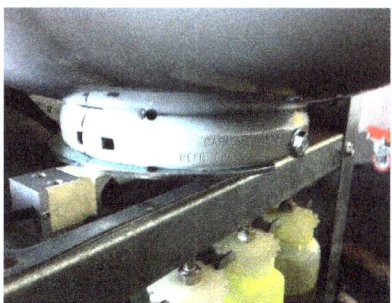	The amount of refrigerant used in an air conditioning system is measured in weight. In order to accurately assess the amount of refrigerant recovered from a system and how much to use for a recharge, the recovery machine needs to be able to weigh the storage container. The storage container will often be mounted on a set of very accurate scales which can measure the weight of the container and any stored gas/refrigerant.
Filter 	As refrigerant is recovered from an air conditioning system it passes through a filter drier unit. This way, the refrigerant in the storage container remains clean and ready to be used in a recharge.
Solenoid valve block 	A series of electrical solenoid operated valves are used to control the flow of refrigerant between different parts of the recovery machine. Differences in pressure between the air conditioning system and the storage container can now be utilised during recovery and filling/charging.
Vacuum pump 	In order to fully empty an air conditioning system of refrigerant, the recovery machine needs to place the system under vacuum. The vacuum pump allows a negative pressure to be created which effectively suctions the refrigerant from the system and lowers the pressures so that any refrigerant dissolved in the lubrication oil 'boils off'.

Refrigerant Handling

Table 1.12 The component parts of an air condition service/recovery machine.

Recovery machine component	Description
Waste oil collection 	During the evacuation of an air conditioning system and the recovery of refrigerant, some of the system lubrication oil is occasionally removed. The recovery machine is able to separate the oil and store it in a see-through container. This way the operator can measure the amount of oil removed and compensate for this when refilling.
New oil container 	This is a container in which the correct grade and quantity of air conditioning system oil can be supplied for injection during filling.
Fluorescent dye container 	To help an air conditioning technician locate the possible source of system leaks, a fluorescent dye can be injected during the refilling procedure. If a system leak is present, its location can often be found by using a UV light to highlight staining on the system components created by the fluorescent dye. If the recovery machine has a method of injecting dye during the recharge process, a container will be used to store the dye until needed.

If an air conditioning leak is suspected, you should not try and charge the system with fresh refrigerant as this might cause environmental pollution. Instead the system should be pressurised with oxygen free nitrogen OFN to help determine the location of the leak.

Refrigerant Handling

Although the outer ends of the hoses on a recovery machine have different sized connectors, the inner ends, where they attach to the machine, are often a standard screw size. This means that it is possible to connect the hoses the wrong way round to the recovery machine and therefore also to the air conditioning system about to be serviced. Always double check the hoses have been correctly attached to the recovery machine and that the connections are properly tightened.

The electronic scales used to measure the weight of the recovery machine storage container and refrigerant are very delicate. If the machine is to be transported any distance, a method is often available which is able to physically lock the scales to prevent any damage or disruption to their accuracy. Always remember to release the scales lock before operating the machine following transportation.

Refrigerant transfer

To transfer refrigerant from a storage cylinder to the recovery machine, or from the recovery machine to the storage container, certain methods should be used.

Transfer refrigerant from cylinder to recovery machine

Step 1
- Switch on recovery machine and allow it to self-calibrate.

Step 2
- Select correct refrigerant container.

Step 3
- Attach a quick service connection adapter to the valve of the refrigerant container.

Step 4
- Wearing appropriate PPE, connect the high pressure hose from the recovery machine to the refrigerant container. Screw in the quick service connector valve, locking the service connector in place and opening the valve.

Step 5
- Open the main container valve and invert the bottle so that it rests on the filler neck stand.

Step 6
- Following the recovery machine instructions, set the amount of refrigerant to be transferred on the keypad, open the high pressure tap and press start.

Refrigerant Handling

Step 7 • Allow the refrigerant to transfer between the cylinder and machine.

Step 8 • When transfer is complete, turn the refrigerant container upright and close the main cylinder valve.

Step 9 • Using the machines recovery procedure, empty the high pressure hose.

Step 10 • Close the machine high pressure tap.

Step 11 • Disconnect the high pressure hose from the container, and correctly store on the recovery machine.

Step 12 • Remove PPE.

When filling the recovery machine from a refrigerant container, it is common for more refrigerant than set to be transferred. This is because the weight of the refrigerant cylinder is not measured using a set of scales. Instead, the recovery machine will often transfer approximately half of the required amount and estimate the rest from what is left over in the hose.

When filling the recovery machine from a refrigerant cylinder, the cylinder must be inverted. This is because there is no **'dipper pipe'** included in the design of the refrigerant cylinder. If left upright during transfer, the recovery machine would attempt to transfer gas. If the cylinder is inverted, it is able to transfer liquid refrigerant.

Dipper pipe - a short length of tube, designed to extend from the main refrigerant cylinder valve to the bottom of the container.

Refrigerant Handling

Transfer refrigerant from recovery machine to cylinder

Step 1
- Switch on recovery machine and allow it to self-calibrate.

Step 2
- Select correct refrigerant container, and ensure that there is adiquate space to allow the correct filling weight.

Step 3
- Attach a quick service connection adapter to the valve of the refrigerant container.

Step 4
- Wearing appropriate PPE, connect the high pressure hose from the recovery machine to the refrigerant container. Screw in the quick service connector valve, locking the service connector in place and opening the valve.

Step 5
- Open the main container valve. (The refrigerant cylinder does not have to be inverted when transferring from the recovery machine).

Step 6
- Following the recovery machine instructions, set the amount of refrigerant to be transferred/injected on the keypad, open the high pressure tap and press start.

Step 7
- Allow the refrigerant to transfer between the machine and cylinder.

Step 8
- When transfer is complete, close the main cylinder valve.

Step 9
- Using the machines recovery procedure, empty the high pressure hose.

Step 10
- Close the machine high pressure tap.

Step 11
- Disconnect the high pressure hose from the container, and correctly store on the recovery machine.

Step 12
- Remove PPE.

Refrigerant Handling

Determining the safe allowable filling weight for a recovery cylinder

Although many refrigerant cylinders may have stickers or labels showing their safe allowable filling weights, an expansion space of 25% must be left for safety. This means that the maximum allowable filling weight is 75% of the total.

To calculate the allowable filling weight, the following formula should be used:

Gross weight - **Tare weight** x 0.75 = maximum allowable filling weight

Gross weight - the total weight of the cylinder when completely filled with refrigerant.

Tare weight - the weight of an empty refrigerant cylinder.

Refrigerant in the high pressure system or hose will be in a liquid state. Refrigerant in the low pressure system or hose will be in a vapour/gaseous state.

Always use recommended recovery procedures to minimise the risk of refrigerant emissions escaping to atmosphere.

Operate a recovery set

Step 1
- Switch on recovery machine and allow it to self-calibrate.

Step 2
- Run vehicle with air conditioning on full for five minutes.

Step 3
- Ensure that you have identified the type of refrigerant contained in the system.

Step 4
- Using technical data, locate the positions of the air conditioning high pressure and low pressure service ports.

Refrigerant Handling

- **Step 5**: Wearing appropriate PPE, connect the recovery machine hoses to the high pressure and low pressure service ports of the vehicle.

- **Step 6**: Following the recovery machines operating instructions, set the recovery and recharge requirements.

- **Step 7**: Set the system vacuum time. A minimum of 30 minutes is recommended to allow for full recovery and **boil off**.

- **Step 8**: Set the vacuum hold time. This will ensure that any system leaks can be located. (If the vacuum decays during the hold period, a system leak is present and must be investigated before attempting a recharge). A minimum of 5 to 10 minutes is recommended to help check for leaks.

- **Step 9**: Set the quantity of system lubrication oil to be injected. (Some recovery machines require that you assess the amount of oil removed during recovery, while others can automatically compensate for any oil that has been removed).

- **Step 10**: Set the amount of tracer dye to be injected. Always follow manufacturer's instructions as excess dye may cause the compressor to **hydro-lock**.

- **Step 11**: Set the required filling quantity of new refrigerant. This is normally measured in weight and can be found in vehicle technical data or on a sticker under the bonnet.

- **Step 12**: Open he high and low pressure taps and start the machine. If the recovery machine is fully automatic, you can now leave this working until the system has been recharged.

- **Step 13**: At the end of the recharge cycle, system operating pressures should be checked.

- **Step 14**: Close the air conditioning service port valves and using the machines recovery procedure, empty the high and low pressure hoses.

- **Step 15**: Disconnect the hoses from the air conditioning system and correctly store on the recovery machine. (Be careful, the high pressure coupler can become extremely hot).

- **Step 16**: Replace protective caps on vehicle's service fittings.

- **Step 17**: Remove PPE

Refrigerant Handling

Boil off - the evaporation of refrigerant that has dissolved in the system lubrication oil. The low pressure created by the vacuum period of the recovery machine causes any dissolved refrigerant to turn into a gas and separate from the oil.

Hydro-lock - a condition caused by the induction of liquid into the compressor. Liquids are virtually incompressible and will cause the pumping mechanism of the compressor to lock solid.

Carry out or carry over - oil that has been drawn out of the air conditioning system during recovery.

Remember that if the pressure gauges read zero on the recovery machine, this doesn't mean that the system is empty, it just means there is no pressure. To ensure that the air conditioning system is empty, it must be placed under vacuum.

Although some air conditioning systems only have a high pressure service port, where possible you should always recover from both high and low systems. This equalising of pressures can often reduce the amount of oil drawn out of the system during recovery, known as **carry out** or **carry over**.

Recording information and making suitable recommendations

At all stages of air conditioning service, maintenance or repair, you should record information and make suitable recommendations. The following table gives examples of how to do this.

Table 1.13 Recording information and making suitable recommendations/giving feedback

Stage	Information	Recommendations
Before you start	Record customer/vehicle details on the job card. Make a note of the customer's repair request and any issues/symptoms. Locate any service or repair history.	Advise the customer how long you will require the car. Describe any legal, environmental or warranty requirements.
During diagnosis and repair	Carry out diagnostic checks and record the results on the job card or as a printout from specialist equipment. List the parts required to conduct a repair.	Inform your supervisor of the required repair procedures so that they can contact the customer and gain authorisation for the work to be conducted.

Refrigerant Handling

Table 1.13 Recording information and making suitable recommendations/giving feedback

When the task is complete	Note down any other non-critical faults found during your diagnosis, service or repair.	Inform the customer if the vehicle will need to be returned for any further work.
	Write a brief description of the work undertaken.	
	Record your time spent and the parts used during the diagnosis, service and repair on the job card. (This information should be as comprehensive as possible, because it will be used to produce the customers invoice).	Advise the customer of any other issues noticed during the repair.
	Complete any service history as required.	

Check your knowledge

1. Which of the following characteristics is true about the Refrigerant R134a?
a R134a is a harmful Greenhouse gas.
b R134a is corrosive to metal.
c R134a depletes the ozone layer.
d R134a does not affect climate change.

2. What type of refrigerant is R1234yf?
a Dichlorodifluoromethane
b Tetrafluoropropene
c Chlorodifluoromethane
d Tetrafluoroethane

3. From which date was the recovery of R12 from a vehicle air conditioning system by MAC operatives banned?
a January 1st 2001
b July 1st 1992
c There are no laws on the refilling of air conditioning systems with R12.
d January 1st 2010

4. An R12 air conditioning system is empty and needs re-gassing; what should the technician do?
a Refill the AC system with R1234yf.
b Convert the AC system from R12 to R134a (retro-fit).
c Refill the AC system with R12.
d Refill the AC system with a blend mixture of R12 and R744.

5. Before fitting a new component of the air conditioning refrigerant system what must be done?
a The technician must wear barrier cream.
b The refrigerant must be safely recovered.
c Nothing.
d All workshop doors and windows should be closed to prevent the escape of refrigerant to atmosphere.

Refrigerant Handling

6. A vehicle air conditioning system that contains R12 refrigerant can sometimes be identified by?
a The diameter size of the flexible hoses.
b The colour of the suction accumulator.
c The high and low pressure quick connection type service connectors.
d The screw type 1/4" SAE flare low pressure service connection.

7. Which item of PPE will the MAC technician need to wear when connecting an air conditioning service machine to a vehicle AC system?
a Goggles.
b Protective clothing.
c Protective gloves made from fluoroelastomer.
d All of the answers.

8. What is the boiling point of R134a at normal atmospheric pressure?
a 100°C
b 26.3°C
c 0.0°C
d -26.3°C

9. Which agreement was designed to reduce the emissions that affect the world's climate?
a G7
b Montreal
c Kyoto
d European

10. What injury could result if refrigerant comes into contact with the skin?
a Scalding of the skin.
b Nothing as the refrigerant will be at normal room temperature.
c Frostbite.
d Dermatitis.

Answers: 1a, 2b, 3a, 4b, 5b, 6d, 7d, 8d, 9c, 10c

Mobile Air Conditioning Principles

Chapter 2 Mobile Air Conditioning Principles

This chapter will help you develop an understanding of the construction and operation of air conditioning components. It shows how these components interact and will help you develop the skills that you need to service these systems. It supports you by providing knowledge that will help you when undertaking both theory and practical assessments. Remember to work safely at all times and observe the relevant environmental, health and safety regulations, while developing air conditioning service routines that are systematic and effective.

Contents

The function and purpose of air conditioning systems	47
Component identification, function and operation	48
Processes, tools & equipment used when maintaining air conditioning systems	58
Basic electrical principles	68
Electrical components used in air conditioning systems	81

Safe working when handling refrigerant

There are many hazards associated with the service and maintenance of air conditioning systems. You should always assess the risks involved with any maintenance or repair routine before you begin and put safety measures in place.
You need to give special consideration to the possibility of:
• The contamination of refrigerant with hydrocarbons
• The exposure of the refrigerant system to heat, which can lead to pressure increase and explosion
You should always use appropriate personal protective equipment (PPE) when you work on these systems. Make sure that your selection of PPE will help protect you from these hazards.

Personal Protective Equipment (PPE)

Table 2.1 PPE required when working on vehicle air conditioning systems

PPE	Recommendations
Overalls 	Overalls provide protection from coming into contact with oils and chemicals.
Gloves 	Fluroelastomer gloves provide protection from fluorinated refrigerants and help protect the hands from frostbite.

Mobile Air Conditioning Principles

Table 2.1 PPE required when working on vehicle air conditioning systems

PPE	Recommendations
Protective footwear 	Safety boots protect the feet from a crush injury and often have oil and chemical resistant soles. Safety boots should have a steel toe-cap and steel mid-sole.
Goggles 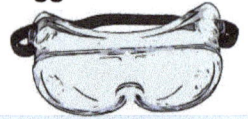	Safety goggles reduce the risk of small objects or refrigerants coming into contact with the eyes.
Bump cap/Hard hat 	A bump cap or hard hat protects the head from bump injuries when working under cars.

Vehicle Protective Equipment (VPE)

To reduce the possibility of damage to the car, always use the appropriate vehicle protection equipment (VPE):

Wing covers

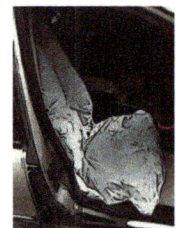
Seat covers

Steering wheel covers

Floor mats

Information sources

The complex nature of air conditioning and climate control systems requires you to have a good source of technical information and data. In order to conduct maintenance and repair procedures, you need to gather as much information as possible before you start.

Sources of information may include:

Table 2.2 Possible information sources

Verbal information from the driver	Vehicle identification numbers
Service and repair history	Warranty information
Vehicle handbook	Technical data manuals
Workshop manuals/Wiring diagrams	Safety recall sheets
Manufacturer specific information	Information bulletins
Technical helplines	Advice from other technicians/colleagues

Mobile Air Conditioning Principles

Table 2.2 Possible information sources

Internet	Parts suppliers/catalogues
Jobcards	Diagnostic trouble codes
Oscilloscope waveforms	On vehicle warning labels/stickers
On vehicle displays	Temperature readings

Always compare the results of any inspection or testing to suitable sources of data. Remember that no matter which information or data source you use, it is important to evaluate how useful and reliable it will be to your safety, maintenance and repair routine.

Air conditioning systems and components

Air conditioning is a convenience system designed to ventilate and lower passenger compartment temperature. It also dehumidifies and purifies the air. A well maintained air conditioning system provides the driver and passengers with an environment which is free from dust and dirt particles, moisture free and will help them to regulate their body temperature. This leads to a comfortable atmosphere in which the drivers and passengers are more alert, helping to improve road safety.

The systems work on the principle of the refrigeration cycle.

- A condensable gas is compressed by a mechanically or electrically driven pump.
- The gas is cooled in a condenser radiator where it changes state from a vapour to a liquid, giving up heat energy as it does so.
- From a storage area, where any water moisture is removed, it passes through a flow control valve.
- The refrigerant passes into an evaporator where its pressure falls and it changes state from a liquid to a gas, absorbing heat energy as it does so.
- From the evaporator, the refrigerant returns to the compressor to begin the cycle all over again.

The refrigerant cycle is a process of heat transfer. It works by absorbing heat energy from the passenger compartment of the car as air is circulated through the evaporator and transferring it to the outside atmosphere as it passes through the condenser radiator on the outside of the car. None of the heat energy is destroyed, it is simply taken from the inside of the car and moved to the outside.

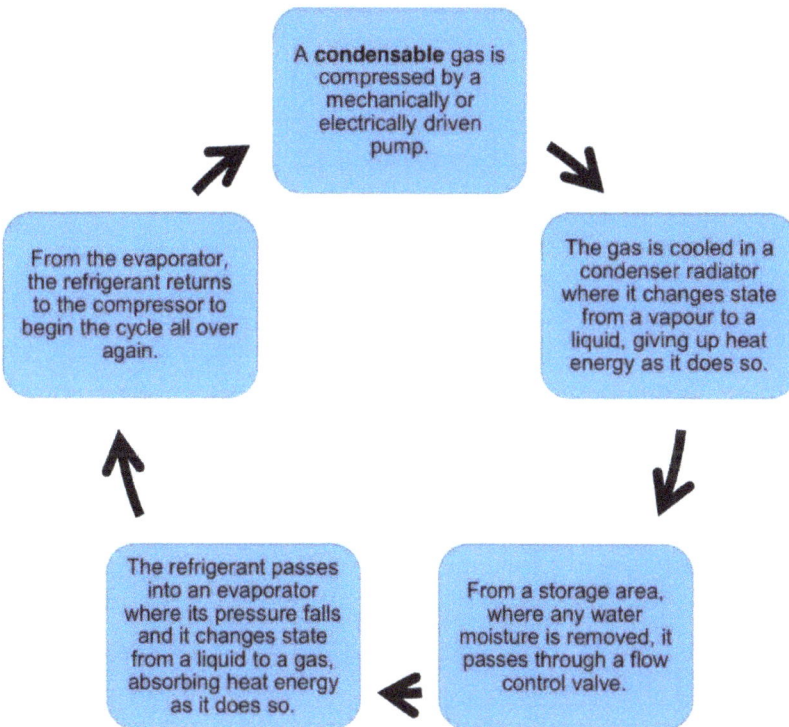

Figure 2.1 The refrigeration cycle

Mobile Air Conditioning Principles

If the evaporator and condenser occupied the same confined space, one would feel cold and the other would feel hot. The refrigeration process could operate continuously, but the overall air temperature would remain the same.

Condensable – an element or a compound that is able to be turned from a vapour/gas to a liquid.

Air conditioning components

Table 2.3 lists the components found in air conditioning systems and an image to help you identify them.

Table 2.3 Identification of air conditioning system components

Air conditioning component and function	Example
Compressor - compresses the refrigerant	
Clutch assembly - disengages the compressor from the drive motor	
Condenser - radiates heat from the refrigerant	
Condenser fan – assists with cooling of the condenser	

Mobile Air Conditioning Principles

Table 2.3 Identification of air conditioning system components

Air conditioning component and function	Example
Evaporator - absorbs heat from the passenger compartment	
Suction accumulator - stores liquid refrigerant before returning to the compressor	
Fixed orifice tube FOT - an accurately sized restriction valve	
Thermal expansion valve TXV - temperature controlled valve	
Receiver drier - stores liquid refrigerant before entering the evaporator and helps to remove any water moisture	
Hoses and pipes - connect various air conditioning components	
Connections - join various air conditioning components	

Mobile Air Conditioning Principles

Table 2.3 Identification of air conditioning system components

Air conditioning component and function	Example
Service ports - allow the safe recovery and filling of refrigerants in an air conditioning system	
Mufflers - act as silencers to reduce any noise created by circulating refrigerant	
Pressure/Temperature switch - control the air conditioning operation due to pressure or temperature	
Relays - control system electric current	

> The air conditioning pressure switch can be combined in one unit to turn off the compressor clutch if system pressures rise too high or fall too low. This is known as a binary switch and is designed to help protect the compressor from excessive loads caused by high or low pressures. High pressure in an air conditioning system may also be an indication of high condenser temperatures. The pressure switch can also sometimes operate the relay controlling the condenser cooling fan if a pre-set pressure value is reached. This type of switch is called a trinary switch as it performs three functions - pressure too high, pressure too low and condenser fan operation.

For a list of refrigerants and their properties, see Chapter 1.

System construction

Air conditioning systems are sealed to prevent the escape of refrigerant gas to atmosphere. They are made using various materials such as:

- Steel - used in the construction of some receiver driers or suction accumulators
- Rubber - used in the construction of system hoses
- Aluminium - used in the construction of condensers and receiver driers

Mobile Air Conditioning Principles

To join the various system components, joints and connections can be:

- Flare type - the end of a metal pipe is spread out to seal against a mating surface when secured by a union nut
- Ring type - the end of a metal pipe is sealed against a mating surface with a rubber 'O' ring and secured with a union nut
- Block/pad type - a metal plate on the end of a pipe or component is squashed against a mating surface by an external clamp; the surfaces will often have a form of rubber gasket to assist with sealing
- Spring lock type - a connection at the end of a pipe can be held against a mating surface by spring pressure; these connections are often quick release if the spring is compressed

To help you identify the different pressure circuits on an air conditioning system, the high pressure section will normally have narrow pipework with a large service connector. The low pressure section will normally have wider pipework with a small service connector.

System layouts and operation

Depending on how the vehicle is to be used, an air conditioning system may have a number of different layouts and configurations. On a standard system, the evaporator is mounded behind the dashboard, just before the heater matrix. Ventilation air can be drawn through the evaporator and heat, moisture and dust/dirt can be removed. The cooler, dry air can then be directed into the passenger compartment to lower the temperature. If the driver requires warm, dry/clean air, the air conditioning system can be adjusted so that having passed through the evaporator the ventilation air is then passed through the heater matrix before entering the passenger compartment. This allows the driver to use the air conditions system all year round to maintain a comfortable environment and also assist with the de-misting of windows during winter months.

Some systems, particularly in vans, may mount the evaporator overhead, allowing the cool clean air to pass down through the passenger compartment. Other systems may have two evaporators to enable different temperatures to be selected (for front and rear passengers for example).

The legal maximum amount of natural leakage of refrigerant, to atmosphere is 40g for a single evaporator system and 60g total for a double evaporator system.

System operation

The refrigeration cycle of an air conditioning system is a closed loop, and as a result has no beginning or end. For the purpose of this description we will start and finish at the compressor.

Fixed orifice tube type (FOT)

1. The compressor pump is operated by the engine or an electric motor. Internal pistons or vanes are used to raise the pressure of the refrigerant gas (compressing the refrigerant will also raise its temperature). This system pressure increase will raise the boiling point of the refrigerant gas.
2. The refrigerant leaves the compressor as a hot high pressure gas.
3. The refrigerant enters the top of the condenser radiator mounted on the outside of the vehicle. As the hot gas passes backwards and forwards across this radiator, air passing through the condenser will transfer some of the heat to the surrounding atmosphere. The fall in temperature of the refrigerant gas, as well as its high

Mobile Air Conditioning Principles

pressure, will allow the refrigerant gas to condense into a high pressure hot liquid (giving up a large amount of latent heat as it changes state).

4. From the condenser radiator, the hot, high pressure liquid refrigerant is transferred via pipes to an accurately sized restriction known as a fixed orifice tube (FOT). The flow of refrigerant to the FOT is controlled by regulating the system pressure, normally by switching the compressor on and off using a clutch mechanism.
5. From the fixed orifice tube, a regulated amount of high pressure liquid refrigerant is allowed to pass through into the evaporator unit, mounted inside the vehicles passenger compartment.
6. The evaporator is located in the low pressure/suction part of the air conditioning circuit. As the high pressure liquid refrigerant enters the low pressure space inside the evaporator, it instantly boils and turns into a low pressure gas (absorbing a large amount of latent heat as it changes state).
7. Passenger compartment air that is passed over the outside of the evaporator by an electrically driven fan has some of its heat energy removed and absorbed into the low pressure refrigerant gas.
8. After the refrigerant leaves the evaporator, it enters a storage container known as a suction accumulator. This contains a silicone desiccant which will remove any water moisture.
9. The low pressure gas (with some heat energy from the passenger compartment) is then drawn back into the compressor where the refrigerant cycle begins all over again.

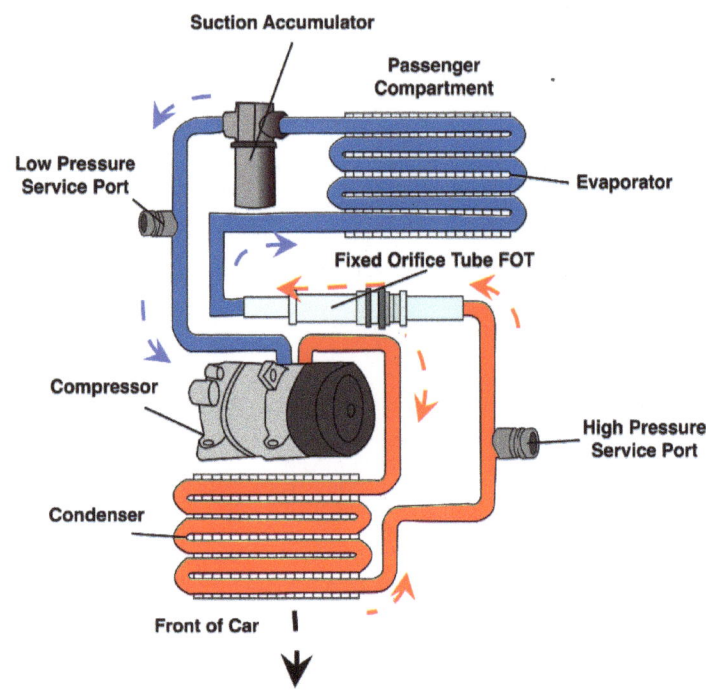

Figure 2.2 FOT air conditioning circuit

A thermal expansion valve system (TXV) follows the same refrigerant cycle but instead of using a fixed orifice tube, it is passed through a temperature regulated valve into the evaporator. To ensure that a sufficient quantity of refrigerant is available for the system, a storage container known as a receiver drier is mounted before the thermal expansion valve. For a full description of the thermal expansion valve TXV system see Chapter 1.

Air distribution

An air conditioning system allows the driver to adjust the levels of ventilation and temperature in the passenger compartment. It also allows the selection of air vents from which the conditioned air is directed. The choice of air direction will often fall into three main areas:

Up - towards the screens and windows

Out - towards the body and face

Down- towards the feet

Mobile Air Conditioning Principles

Ventilation motors

When a vehicle is moving, ventilation air can be achieved using the ram-air action caused by the design of the air intake and the forward motion of the car. So that a constant supply of ventilation air can be achieved even when the vehicle is stationary, air conditioning systems are equipped with an electric motor driven ventilation fan. The speed of the motor can be controlled by the driver by varying the amount of power the electric motor receives. This may be done using voltage dropping resistors, known as rheostats, or through ECU control using **duty cycle**.

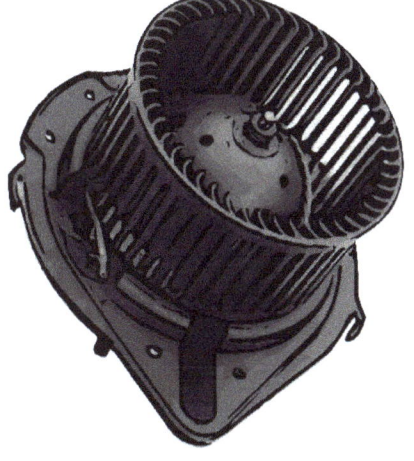

Figure 2.3 A ventilation fan and motor

Duty cycle - the process of switching an electrical component on and off very quickly to control the power provided/consumed.

Convection currents - the transfer of heat energy through a liquid or a gas.

Due to **convection currents,** the quickest way to warm up the passenger compartment is to direct the hot air downwards, and the heat will cause it to rise up. The quickest way to cool down the passenger compartment is to direct the cold air upwards, and convection currents will cause the cooler air to fall.

Controls

The driver will often have a series of controls which enable the setting of the desired temperature and ventilation, some of these are explained in table 2.4.

Table 2.4 Air conditioning controls

Control	Use
Temperature dial 	Normally supplied as a rotational dial, this control allows the driver to regulate the system temperature between hot and cold air. Valves or flaps are used to control the flow of hot or cold air to the passenger compartment. If the driver operates the air conditioning at the same time as a hot temperature is selected, then a blend of conditioned and hot air is supplied.

Mobile Air Conditioning Principles

Table 2.4 Air conditioning controls

Control	Use
Fan speed switch 	Normally supplied as a rotational or slide type switch, this allows the driver to select the speed of the electric ventilation fan. It will usually have three or four speeds and an off position.
A/C switch 	With an air conditioning system the driver is often supplied with a method to switch the refrigerant cycle on and off. This gives them the opportunity to decide if full air conditioning is needed. Switching off the air conditioning when not required can improve vehicle performance, fuel economy and help reduce exhaust emissions.
Recirculation switch 	Much of the air used for ventilation enters the passenger compartment from outside, through vents located near the lower windscreen scuttle. A recirculation switch allows the driver to close these scuttle vents and simply recycle the air in the passenger compartment through the air conditioning system and ventilation. This this can be useful when fumes are present outside the car and reduces unpleasant smells.
Vent selector 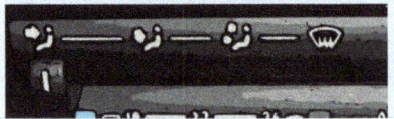	These selectors allow the driver to choose which vents receive the ventilation air. A control operates cables, vacuum valves or electric motors which are able to move flaps in the air conditioning system, directing air to the chosen vents. The option to use various combinations such as windscreen and face vents is often provided.
Vent direction 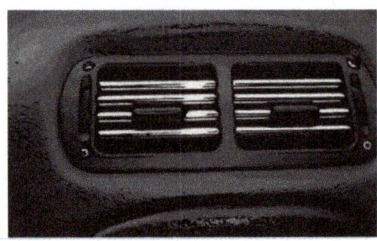	Many vents that are mounted in the dashboard of a vehicle have controls which allow their positions and angle to be altered. This way the driver can select in which direction these vents point.

When the air conditioning is operating on full, it may take around 5 **horse power** from the engine to drive the refrigerant compressor. Many manufacturers will now produce vehicles where the air conditioning is switched on by default, but a manual override will be available to enable the driver to switch the system off.

Mobile Air Conditioning Principles

Horse power - a unit of measurement used to describe the power produced by an engine. 1 Horse power is equal to 746 watts (UK).

Pollen filters

To ensure that the ventilation air entering the passenger compartment is as clean as possible, many vehicles are fitted with pollen filters. The pollen filter is mounted in the inlet of the vehicles ventilation system. It is a pleated micro-fibre material, designed to capture many of the airborne particles such as dust and pollen before they have the opportunity to enter the car. Some pollen filters are impregnated with active charcoal to help reduce the amount of fumes and smells entering the car; these are known as cabin filters. The efficiency of an air conditioning system can be greatly improved if pollen filters are kept in good condition and regularly replaced.

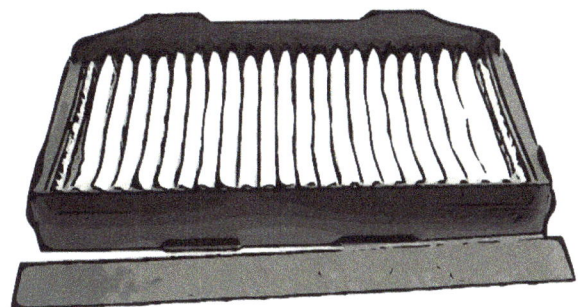

Figure 2.4 A pollen filter

As air is circulated through the air conditioning ventilation system, water held in suspension condenses as it comes into contact with the cold evaporator. This dehumidifies the air, but also helps to clean and remove dust, dirt and smoke particles as they stick to the damp surface of the evaporator. As the system continues to operate, this dust and dirt should drip from a drain tube under the car, along with the condensed moisture. If the system was not fitted with an adequate pollen or cabin filter, the evaporator would quickly become very dirty and the build-up of dirt in a dark damp environment may lead to an issue known as sick car syndrome SCS.

Sick car syndrome

The main cause of sick car syndrome is the result of condensation collecting on a vehicles evaporator. The moist environment tends to harbour fungi, mould and mildew, which grow within the air conditioning system on the core of the evaporator and any foam type insulation. As air passes over the evaporator it picks up these microbes or their waste and then enters the passenger compartment through the vents.

These microbes are a potential source of various ailments and flu like symptoms that the passengers may exhibit including:

- sneezing
- wheezing
- coughing
- eye and mucous membrane irritation
- drowsiness
- chest tightening

Mildew-like odours are released when the air conditioning system is operated following a rest period and in some cases, the odour seems to disappear after a few seconds or minutes of operation. The odour disappearance is caused by the passenger's noses becoming accustomed to the smell -- not that the microbes have disappeared.

Some drivers place air freshener's in their cars to help mask unwanted smells, but this does not combat the bacterial growth on the evaporator. In order to correct this issue, the evaporator must be cleaned. This can be achieved with specialist cleaning agents which are sprayed onto the evaporator or drawn through the air conditioning system vents when the system is operated on recycled air.

Mobile Air Conditioning Principles

In order to deal with smells and sick car syndrome you should:

- ✓ Make sure where the problem is - If the car has an odour when you enter it, the problem is inside the car and not smells from the surrounding air. If you only get an odour when the ventilation fan is on, the problem is most likely in the air conditioning system.
- ✓ Clean the air intake - Remove any dead leaves or plant material from the lower windscreen scuttle or air intake system.
- ✓ Clean up any spills - Dry any water, milk or liquid spills inside the car.
- ✓ Remove rubbish - Check for any stale food inside the car.
- ✓ Valet the interior - After shampooing the car's carpet, dry it thoroughly before you close the windows.
- ✓ Assess any water ingress to the car - Check the luggage area for any water leaks including the spare tyre well.
- ✓ Clean the evaporator - Use a specially formulated evaporator cleaner to disinfect the unit. Access can often be gained to the evaporator when the pollen filter is removed.

The large evaporators used in the air conditioning systems of buildings also suffer with the build-up of mould and bacteria. When operated microbes can be circulated and cause severe illness such as Legionnaires Disease. Legionnaires Disease causes high fever and pneumonia and is potentially fatal.

Pollen sensors

A Micro Dust/Pollen Sensor Unit is a sensing device used to detect the amount of particles in the air, such as pollen, dust, smoke and fur of animals, by using an optical system. Ventilation air is passed through the unit and it operates on the principle of light scattering method. The sensor is designed to detect particles with a size over 1 micro-meter, which means it can detect cigarette smoke, pollen and dust. The Micro Dust/Pollen Sensor Unit detects the amount of particles per unit volume and can be used to provide a warning that the cabin filter should be replaced.

Lubrication

Because the compressor of an air conditioning system is a form of mechanical pump driven from the engine or an electric motor, it requires lubrication in order to prevent the moving metal components rubbing against each other, creating friction, heat and wear. The lubrication oil must be compatible with the refrigerants used and not affect system operation. For this purpose, special oil blends of either Polyalkylene Glycol (PAG) or ester oil are used. There are three main viscosities of PAG oil - 46, 100 and 150, and the quantity and type is recommended by the vehicle/air conditioning manufacturer. PAG oil is hygroscopic, meaning that it absorbs moisture from the atmosphere, because of this, you should only use oil from a sealed container. Water in an air conditioning system will freeze and cause blockages. Ester oil on the other hand is one viscosity and universal, meaning that it should be safe to use on most systems.

Figure 2.5 PAG or ester oil

Mobile Air Conditioning Principles

If an air conditioning system has been retrofitted from R12 to R134a, PAG oil can react with small amounts of chlorine left over from the R12 to cause corrosive mixtures. The system and components must be thoroughly flushed before the PAG oil is added.

Hybrid and electric vehicles use a compressor driven by an electric motor, powered from the high voltage system. These compressors require a special PAG oil in order to function safely. The PAG oil should not readily conduct electricity, as this may cause a short circuit in the high voltage system, making metal components become 'live'. Once opened the hygroscopic nature of the oil means that any moisture absorbed will increase its electrical conductivity; any oil left over should be disposed of following safety and environmental requirements.

Figure 2.6 A high voltage electric powered compressor

Units of measurement

To make sure that manufacturing can be standardised across all countries, there are special organisations which agree on the standards for measuring.

• The British Standards Institution (BSI) is the recognised authority for the measuring standards in the UK. For more information visit the BSI website.

• Control of measuring systems internationally is carried out by the International Organization for Standardization (ISO). For more information visit the ISO website.

The International System of Units (known as the SI system) exists so that all countries use the same measuring standard. This means that each unit of measurement will be exactly the same whatever the country you are in; so a metre is the same whether you are in the UK or any other part of the world.

Although the measurements for many systems use standard values, the exception to this is air conditioning, heating and ventilation where older measurement units are often used which may require conversion to the agreed standard.

Table 2.5 gives some examples of standard units of measurement, with some alternative values used in air conditioning shown. For more information visit the website of the Bureau International des Poids et Mesures (BIPM).

Mobile Air Conditioning Principles

Table 2.5 Some examples of worldwide standard units of measurement with additional air conditioning values

Quality	Standard and Alternative	Abbreviation
Length	Meter (inches also used when describing service port connections)	m (")
Mass/Weight	Kilogram (may also be described as pounds or ounces when measuring refrigerant and PAG oil filling weights)	Kg (lbs or oz)
Pressure	Pascal (may also be described as Barometric pressure or Pounds Per-square Inch when used for positive pressures and inches or millimetres of mercury when describing negative pressures)	Pa (Bar, Psi, InHg or mmHg)
Temperature	Kelvin (often described as degrees Centigrade/Celsius or degrees Fahrenheit with air conditioning)	°K (°C or °F)
Volume	Litre (may also be described as cubic centimetres when measuring PAG oil or dye quantities)	l (cc)
Energy	Joule (may also be described as British Thermal Units when discussing thermal energy quantities)	J (BTU)

The advantages of using set standard measurements are:

• Component parts can be changed easily – all parts are made to the same size by all manufacturers.

• Lower manufacturing costs – manufacturers can share the production costs for different components to make one unit.

• Lower costs to the consumer – parts can be bought from other manufacturers rather than just one, which means prices will have to be competitive.

- Improved quality – all manufacturers will need to stick to specific sizes.
- Servicing and maintenance - standard values can be used to create fully automatic air conditioning recovery management stations.

Processes, tools and equipment used for the servicing and maintenance of air conditioning systems

Many maintenance and repair procedures used for air conditioning systems are undertaken using a recovery management station RMS. Fully automatic recovery means that there is less risk of accidental leakage of refrigerant to atmosphere, and the technician is able conduct other work while a recovery and recharge process is underway. For a description of the components that make up a recovery management station, see Chapter 1 Table 1.12.

Mobile Air Conditioning Principles

Although recovery management stations are now commonly used, individual pieces of equipment can be very effective for specific tasks.

Manifold gauges, hoses and connectors

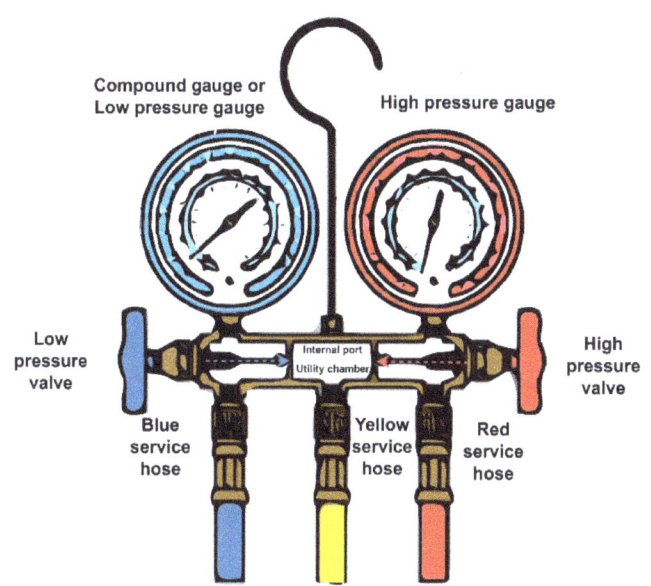

Figure 2.7 Air conditioning manifold gauge set

This piece of equipment consists of a pair of gauges, joined to a common **manifold** with hose attachments to connect it to an air conditioning system and a charging cylinder. A set of valves can then be used to control the flow of refrigerant through the system and the gauges allow pressures to be displayed.

In figure 2.7 the right side of the manifold with the red hand valve is the high side, and the red gauge is the high pressure gauge. The left side of the manifold with the blue hand valve is the low side, and the blue gauge is the vacuum/pressure gauge. Separate ports and passages from the low and high side service hoses to their respective gauges allow the operator to check pressure and vacuum readings when the hand valves are in the open or closed position.

The manifold chamber connects three lower R134a fittings to each other by internal passages.

A calibration screw will be provided to ensure that the gauges are zeroed before the manifold set is connected to an air conditioning system or charging cylinder and pressure is applied. If the gauges do not read 0 psi, you can remove the gauge face and adjust calibration screw to set the needles at 0 psi.

Do not over-tighten hoses on to manifold gauge set.

Never open the high side hand valve on manifold gauge set when the air conditioning system is operating. The high pressures created by the air conditioning compressor may force refrigerant back through the manifold set causing damage to both the A/C system and equipment.

Manifold - a pipe or chamber branching into several openings.

Vacuum pump

An electrically powered vacuum pump can be used to lower air conditioning system pressures to a point where refrigerant dissolved in lubrication oil will change state and boil out. This will ensure correct system evacuation and recovery. A vacuum pump can be designed to evacuate an air conditioning system by drawing a vacuum, or it may also have a discharge port that can be connected to a refrigerant container so that the system can be recovered.

Figure 2.8 Vacuum pump

Mobile Air Conditioning Principles

Weighing scales

The quantity of refrigerant used in an air conditioning system is normally measured by its weight. A very accurate set of digital weighing scales can be placed on a level surface underneath a charging cylinder, and once its initial weight has been measured, this figure can be used to calculate and add the correct amount of refrigerant to a system.

Figure 2.9 Weighing scales

Charging cylinder

The charging cylinder is a metal container that can be pressurised with a single type of refrigerant. Once pressurised, the refrigerant is held safely in a liquid state. The charging cylinder will be colour coded to indicate its contents, and should display information about dangers, precautions and allowable filling weights. For more information on charging cylinders, see Chapter 1.

Figure 2.10 Charging cylinder

Oil injection equipment

Figure 2.11 Oil injection equipment

Oil injection equipment can be used to add lubrication oil to an air conditioning system if components are replaced or if waste oil has been drawn out during the system evacuation process. The oil injector is normally a graduated container that can be opened up and filled with the correct quantity and grade of air conditioning oil, then closed and sealed. It will have two service connectors so that it can be joined in series with the air conditioning equipment or manifold gauges. A valve mechanism is often also included in the design so that the flow of oil into the air conditioning system can be regulated. If the system has been recently recovered, the vacuum that exists from the evacuation process can be used to draw a measured quantity of oil into the air conditioning circuit. If components have been replaced or the system is just being refilled, pressure from the refrigerant filling process can be used to inject a measured quantity of oil into the air conditioning circuit.

> Always follow manufacturer's recommendations for the correct type and quantity of oil to use when adding to an air conditioning system.
>
> Although ester oil is often considered a universal substitute for PAG oil, this may not be suitable for some systems, especially for use in hybrid and electric vehicles. If the incorrect oil is used it may cause system damage or in the case of hybrid and electric vehicles, allow the conduction of high voltage electricity to vehicle components. If system components become electrically conductive, this may lead to fire, electric shock, injury or death.

Mobile Air Conditioning Principles

Leak detection equipment

If an air conditioning system has a small leak, this can often be very difficult to find. A number of different types of leak detection equipment are available to help you diagnose and repair faults.

- Fluorescent dye and ultra-violet light - in this design of leak detection, a dye is introduced into the air conditioning system along with the refrigerant. This can be done manually through an open pipe using an injection gun or using a system similar to that of the oil injection described earlier. Once the air conditioning system has been operated, any leaks will be highlighted by the dye. If an ultra-violet light and yellow tinted safety glasses are also used, the dye will glow, making detection easier.
- Electronic detection - in this design, a portable handheld detection unit is used to sense small quantities of leaking refrigerant. When held near the location of a suspected leak, a sensor on the unit is able to detect refrigerant vapour (measured in parts per million ppm) and illuminate LED's or sound a buzzer. Ultrasonic electronic detectors are also available, which when placed near the source of a suspected leak can detect the fault by the sound being produced. An ultrasonic detector uses an electronic process called "heterodyning" to convert the high frequency leak sound to a lower range where the hissing of the leak can be heard through a set of headphones, and traced to its source. Any turbulent gas will generate ultrasound when it leaks, therefore it does not matter what refrigerant you are leak testing, including the use of oxygen free nitrogen.
- Halide detection - this is an old fashioned type of refrigerant detection. A small canister of butane or propane gas, provides the source of fuel for small burner unit, similar in construction to a blow-torch. When lit, the burner should produce a clean blue flame. When held near the location of a suspected leak, the flame will change colour if it comes into contact with leaking refrigerant gas.

Figure 2.12 UV leak detection kit

Halide leak detection is now rarely used due to the risks this type of equipment presents. The naked flame is able to provide a source of ignition for any flammable materials in the immediate vicinity, and when the flame contacts refrigerant gas this may be converted by thermal decomposition into harmful substances (see Chapter 1).

Oxygen free nitrogen OFN

Anyone wishing to service and repair air conditioning systems should have access to oxygen free nitrogen OFN. This is because knowingly charging a leaking system with refrigerant and allowing it to escape to atmosphere is illegal. OFN is an inert gas that can be used to pressurise an air conditioning system and help diagnose leaks because it has no harmful effects on the environment. A gas cylinder containing OFN can be connected to the air conditioning system via a high pressure pipe and quick fit connector. The system should be pressurised to around 10 Bar and an inspection conducted to see if any leaks are present. Once the system has been placed under pressure, it can be checked for leaks using fluorescent dye or ultrasonic detection.

Figure 2.13 Oxygen free nitrogen OFN

Mobile Air Conditioning Principles

A solution of soapy water can sometimes be used to help locate an air conditioning leak from a connection or joint. When applied to connections or joints, the solution will begin to bubble if a moderate or severe leak exists.

Digital thermometers

In order to assess the efficiency of an air conditioning system, temperature at various points needs to be measured. Digital thermometers are used for this process as they are easy to read, accurate and sensitive to small amounts of temperature change; updating the displayed values very quickly. Non-contact digital thermometers are available that allow the technician to quickly scan many air conditioning system components and check their function.

Figure 2.14 Non-contact digital thermometer

Refrigerant identifier

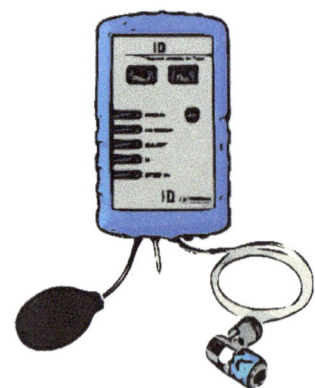

A refrigerant identifier is often a handheld device that can be connected to an air conditioning system or charging cylinder to help the operator determine the type of gas contained. Depending on the equipment type, a high or low pressure service port connector is attached to the device and a sample of gas is drawn into the tool by system pressure or a small hand pump. An inferred sensor inside the device is then able to detect the gas type and will display the results on a screen or through LED's. Some refrigerant identifiers are able to show the percentage of gas type and will also sound an alarm if hydrocarbon gasses are present in the system or cylinder. Using a refrigerant identifier helps to prevent **cross contamination** of gasses, protecting systems, components and recovery equipment.

Figure 2.15 Refrigerant identifier

Air is a non-condensable gas. If air is present in an air conditioning circuit, it will not take part in the refrigeration process, reducing the overall effectiveness of the system. Some more advanced refrigerant identifier tools are able to safely release this air from the system in a process known as "**purging**".

Cross contamination - the accidental mixing of refrigerants and other gases in an air conditioning system or cylinder.

Purging - the safe release of trapped air from an air conditioning system.

Mobile Air Conditioning Principles

Refrigerant flushing equipment

Air conditioning system components may require flushing if they have become contaminated with excessive oil or blockages have been produced due to a build-up of dirt and debris. Some components can be successfully flushed while others will require replacement. System flushing should only be conducted if you have had adequate training.

Some vehicle manufacturers do not recommend flushing air conditioning components with solvents. Doing so may cause damage or invalidate manufacturers' warranties. Always check manufacturer's recommendations before flushing any system components.

Air conditioning components that should **not** be flushed include:

- The compressor
- Receiver driers
- Suction accumulators
- Thermal expansion valves
- Fixed orifice tubes
- Any component containing a muffler

These components may trap the solvent flush, causing damage to the system and compressor when reassembled and operated. Potential things to consider when choosing a solvent for use in flushing air conditioning system components include:

- Is it safe for use in automotive systems
- How easy is it to use and remove
- What dangers are involved with its use (i.e. flammability)

You should **never** use:
- Mineral based solvents
- Alcohol
- Carburettor cleaner
- Brake cleaner
- Petrol
- Unfiltered compressed air

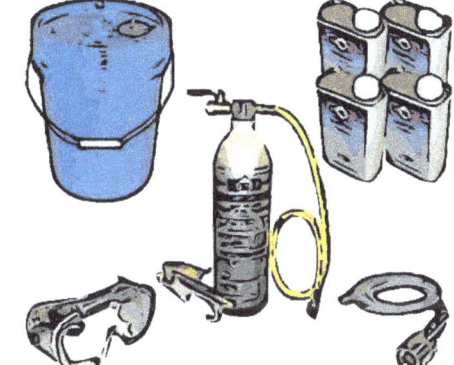

Figure 2.16 Refrigerant flushing equipment

Two methods that can be used to flush a system with a recommended solvent are described in the next section.

Flushing bottle and spray nozzle - this tool is a container with attachments that allow it to be connected to the removed air conditioning component, and a regulating valve that allows it to be connected to compressed air. Once the flushing bottle has been pressurised, it is connected to the required component, and the solvent flush is allowed to flow into the system part being cleaned. Once the solvent has been introduced and given time to clean, the component must then be purged with filtered compressed air at around 80 psi for a minimum of 30 minutes before reassembly. Excess dirty solvent can be collected using a return pipe and solvent proof container.

Mobile Air Conditioning Principles

If the solvent used to flush an air conditioning component is not fully purged (i.e. by using filtered compressed air for a minimum of 30 minutes, allowing for correct evaporation) damage will be caused to the compressor when the system is reassembled and operated.

Any dirty/waste flushing solvent must be disposed of correctly, following all environmental and safety procedures. See controlled waste section in Chapter 1.

No matter which type of flushing equipment is used, always follow manufacturer's instructions.

Automatic flushing machine - this piece of equipment can conduct the cleaning of air conditioning system components automatically. Once the component to be cleaned has been disconnected from the system, the flushing machine is connected using the provided adapters. Workshop compressed air is then connected to the flushing machine and once the valves are opened, the component can be cleaned and purged automatically.

When using an automatic flushing machine, the dirty solvent leaving the component is often passed through a filter and then reintroduced over again to improve the cleaning process.

Compressor drive belt tensioners

One component that is often overlooked when maintaining and servicing a light vehicle air conditioning system is the auxiliary drive belt which is used to operate an engine driven compressor. This should be inspected for condition and tension.

Although an experienced technician can often assess belt tension by feel, a number of tools are available which can assist in the measurement of correct belt tension.

Deflection gauges - this is a tool that is placed on the longest run of the auxiliary drive belt and has a knob which is turned. A plunger attached to the adjustable knob is then able to operate a gauge showing the actual tension of the belt. This reading can then be compared to a list of acceptable tolerances provided by the tool manufacturer and adjusted until it is an acceptable value.

Harmonic tension - this uses a small handheld electronic device, about the size of a multimeter, to 'listen' to the tension of the auxiliary drive belt. A microphone is first calibrated following manufacturers procedures, and then it is held close to the drive belt near the longest run accessible. The belt can then be flicked with your finger, and the sound emitted will be picked up by the microphone. This sound is converted into a frequency measurement which is then displayed on a small LCD screen (the higher the frequency, the tighter the belt). This frequency measurement can then be compared with a table supplied by the tool manufacturer to see if the tension is within acceptable limits.

Figure 2.17 Auxiliary belt tension gauge

Mobile Air Conditioning Principles

Servicing procedures

The following service procedures are for individual tools. For service procedures using a recovery management station (RMS) see Chapter 1.

System recovery

A vacuum pump can be used to lower air conditioning system pressures for evacuation. In order to safely recover refrigerant from an A/C system, the vacuum pump must be equipped with a discharge port that can be connected to a recovery cylinder. The following procedure covers system recovery.

Evacuating/recovery an air conditioning system with individual tools

Step 1
- Observing all health and safety, including PPE, inspect tools, equipment and charging cylinders for damage. (Ensure that all valves on the manifold set are closed).

Step 2
- Connect the manifold gauge set blue low side service hose to the air conditioning systems low pressure service port.

Step 3
- Connect the manifold gauge set red high side service hose to the air conditioning systems high pressure service port.

Step 4
- Connect the centre yellow charging hose to a vacuum pump, and the discharge hose from the vacuum pump to a suitable recovery cylinder, ensuring you have checked/calculated allowable filling weights.

Step 5
- Open the high and low manifold valves using the red and blue hand valves and start the vacuum pump. Open the valve on the recovery cylinder and operate vacuum for a minimum of 30 minutes.

Step 6
- After evacuating and recovering the system according to the manufacturer's specifications, close both the high and low side hand valves and turn off the vacuum pump.

Step 7
- Close the valve on the recovery cylinder and disconnect all hoses. (Be careful, the high pressure coupler can become extremely hot).

Step 8
- Replace protective caps on vehicle's service fittings.

Always connect hoses to the manifold gauge set and R134a quick couplers before connecting R134a quick couplers to the R134a service ports on the A/C system. This will help to prevent refrigerant escaping from the open fittings and prevent any air or refrigerant entering an evacuated system.

Never open the red high pressure hand valve when the A/C System is operating.

Charge with gas only, never with liquid refrigerant.

Mobile Air Conditioning Principles

Charging an air conditioning system with individual tools

Step 1 • Vehicle engine must be not running and the A/C system must be turned off. If possible, use a vacuum pump to evacuate the system for a minimum of 30 minutes before charging.

Step 2 • Make certain that the valve on the refrigerant cylinder is in the closed position.

Step 3 • Connect the manifold gauge set blue low side service hose to the air conditioning systems low pressure service port.

Step 4 • Connect the manifold gauge set red high side service hose to the air conditioning systems high pressure service port.

Step 5 • Connect the centre yellow charging hose to the charging cylinder and open the cylinder valve, but do not invert the cylinder.

Step 6 • Open the low side valve on the manifold gauge set.

Step 7 • Charge system until gauges read just over 50 psi of pressure.

Step 8 • Close low side valve on the manifold gauge set.

Step 9 • Both high side and low side valves on the manifold gauge set must be turned off.

Step 10 • Start the vehicle, ensuring adequate air flow around the condenser and radiator to avoid overheating.

Step 11 • Turn on the air conditioning and set the controls for maximum cooling and high fan speed.

Step 12 • Open the valve on the refrigerant charging container. Keep the refrigerant charging container upright at all times so that refrigerant can only enter the hoses as a gas.

Mobile Air Conditioning Principles

Step 13 • Using a set of electronic weighing scales placed below the refrigerant charging container, allow gas to flow into the system until the desired amount has been added according to manufacturer's specifications. (Getting the required refrigerant amount into the system may take several minutes).

Step 14 • When you have charged the system to manufacturer's specifications, close the manifold's blue low side hand valve. Let the compressor run and check the gauge readings to be sure the system is operating properly.

Step 15 • Close the valve on the charging cylinder and disconnect all hoses. (Be careful, the high pressure coupler can become extremely hot).

Step 16 • Replace protective caps on vehicle's service fittings.

Using a manifold gauge set to check air conditioning system pressures

Step 1 • Observing all health and safety, including PPE, inspect tools and equipment for damage. (Ensure that all valves on the manifold set are closed).

Step 2 • Connect the manifold gauge set blue low side service hose to the air conditioning systems low pressure service port.

Step 3 • Connect the manifold gauge set red high side service hose to the air conditioning systems high pressure service port.

Step 4 • Start the vehicle, then turn on the A/C system, and let it run until the gauge readings stabilise. Compare readings with manufacturer's specifications.

Step 5 • If the pressure shown on the gauges differ from the manufacturer's specifications, determine the problem and make the necessary repairs (see Chapter 3).

Step 6 • If the gauge readings are within the manufacturer's specification, and the A/C system appears to be operating properly, stop the A/C system, turn off the vehicle, and disconnect the hoses from the system. (Be careful, the high pressure coupler can become extremely hot).

Step 7 • Replace protective caps on vehicle's service fittings.

Mobile Air Conditioning Principles

Electrical principles

The operation and control of vehicle air conditioning systems is achieved using electrics and electronics. This section will help you develop an understanding of the electrical and electronic principles behind these systems as well as their construction.

When working with light vehicle electrical and electronic systems, the main hazard is the possible risk of electric shock. Although most light vehicle electrical systems operate with low voltages of around 12V, an accidental electrical discharge can be caused by incorrect circuit connection. This can be enough to cause severe burns.

When working on the air conditioning systems of hybrid and all electric vehicles, take care around the high voltage components. The high voltage system can normally be identified by its reinforced insulation and shielding which is often coloured bright orange. These systems carry voltages that can cause severe injury or death.

Some electrical components, such as car batteries, contain dangerous chemicals or acids. If these are allowed to come into contact with your skin, they can cause injury. Your selection of personal protective equipment (PPE) and your manner of working when you work on electrical systems should protect you from these hazards.

What is electricity?

Every substance known to man is made of molecules. The molecules of a substance are made up from **atoms**. For example, if the substance is water, the molecule is H2O. This means that the molecule is made up of two hydrogen (H) atoms joined to one oxygen (O) atom.

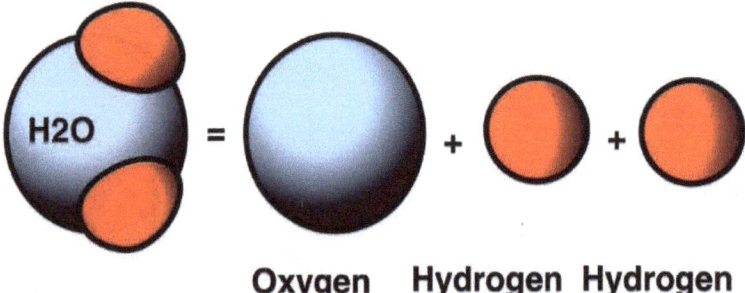

Figure 2.18 The atoms in water

The reason why it can be difficult to understand electricity is because it is contained within atoms. Atoms are very small and hard to imagine.

• The easiest way to imagine an atom is like a miniature solar system, with a sun in the middle and planets orbiting around the outside.

• In the case of an atom, the nucleus represents the sun. The **nucleus** is made of positively charged particles known as **protons**. It also contains particles with no charge known as **neutrons**.

Mobile Air Conditioning Principles

- Orbiting around this nucleus (in a similar way to the planets) are negatively charged particles known as **electrons**. As the name suggests, it is the electrons that produce electric current.

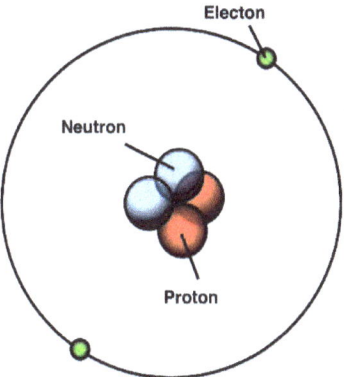

Figure 2.19 An atom

Different atoms have different numbers of protons and electrons, as shown in the periodic table.

1 H																	12 He
3 Li	4 Be											5 B	6 C	7 N	8 O	9 F	10 Ne
11 Na	12 Mg											13 Al	14 Si	15 P	16 S	17 Cl	18 Ar
19 K	20 Ca	21 Sc	22 Ti	23 V	24 Cr	25 Mn	26 Fe	27 Co	28 Ni	29 Cu	30 Zn	31 Ga	132 Ge	33 As	34 Se	35 Br	36 Kr
37 Rb	38 Sr	39 Y	40 Zr	41 Nb	42 Mo	43 Tc	44 Ru	45 Rh	46 Pd	47 Ag	48 Cd	49 In	50 Sn	51 Sb	52 Te	53 I	54 Xe
55 Cs	56 Ba	57-71	72 Hf	73 Ta	74 W	75 Re	76 Os	77 Ir	78 Pt	79 Au	80 Hg	81 Tl	82 Pb	83 Bi	84 Po	85 At	86 Rn
87 Fr	88 Ra	89-103	104 Rf	105 Db	106 Sg	107 Bh	108 Hs	109 Mt	110 Ds	111 Rg	112 Cn	113 Uut	114 Fl	115 Uup	116 Lv	117 Uus	118 Uuo

57 La	58 Ce	59 Pr	60 Nd	61 Pm	62 Sm	63 Eu	64 Gd	65 Tb	66 Dy	67 Ho	68 Er	69 Tm	70 Yb	71 Lu
89 Ac	90 Th	91 Pa	92 U	93 Np	94 Pu	95 Am	96 Cm	97 Bk	98 Cf	00 Es	100 Fm	101 Md	102 No	103 Lr

Figure 2.20 The periodic table of elements

Movement of electrons

To make the electric current, you need to move electrons from one atom to the next. To do this they need to be given a push by an external force or pressure.

The pressure used to move electrons can be created by:

- Magnets
- A chemical reaction

Mobile Air Conditioning Principles

Orbiting electrons are held in place in a similar way to the gravity acting on the planets circling around the sun. Because the hydrogen atom is so simple, the attraction between the nucleus and the electron is very strong. This makes it very hard to move the electron. When electrons don't move easily the element is known as an **insulator**.

A copper atom contains 29 electrons and 29 protons. The electrons orbit in circles that get bigger and bigger. The electrons in the farthest orbit have a far weaker bond/attraction to the nucleus than those in a hydrogen atom. These outer electrons are known as 'free electrons'. If an external pressure is applied, electrons can be moved from one atom to the next. This movement of electrons is electric current. When electrons do move easily the element is known as a **conductor**.

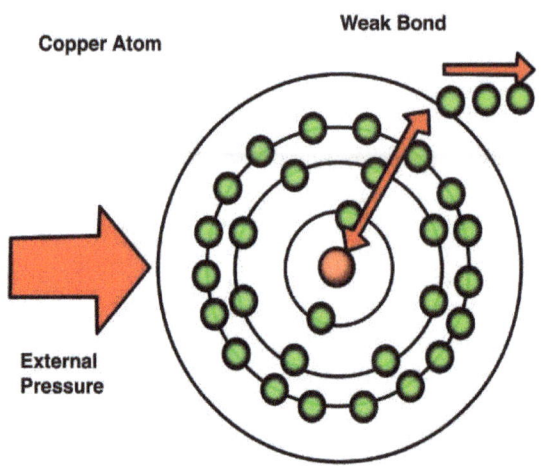

Figure 2.21 A copper atom with external pressure applied

Atom – the smallest component of any chemical element.

Nucleus – the centre of an atom.

Protons – positively charged particles.

Neutrons – particles with no charge.

Electrons – negatively charged particles.

Insulator – a chemical element which does not allow the easy movement of electrons.

Conductor – a chemical element that allows the easy movement of electrons.

Electric current

When electricity flows or moves, this is known as '**current**'. There are two types of current – **alternating current (AC)** and **direct current (DC)**.

Current – moving electricity.

Alternating current (AC) – electricity that moves in two directions (backwards and forwards).

Direct current (DC) – electricity that only moves in one direction.

Mobile Air Conditioning Principles

The direction of current flow and electron flow

Electric current will always move from an area of high pressure to an area of low pressure. In conventional or standard electrics, the positive (+) side of the circuit has high pressure, and the negative (–) side of the circuit has low pressure. This means that electricity leaves the battery at the positive terminal, flows through the circuit and then re-enters the battery at the negative terminal. Whenever you test an electric circuit on a vehicle you should use conventional electrics theory (follow the circuit from positive to negative).

True electron flow is in the opposite direction to standard or conventional current (negative to positive). Remember this if the words 'electron flow' are ever used in a description or test.

Conductors and insulators

- Conductors are used on cars where we want electricity to flow easily, such as wiring.

- Insulators are used on cars to reduce the movement of electricity, such as the coating on the outside of a wire.

Circuits

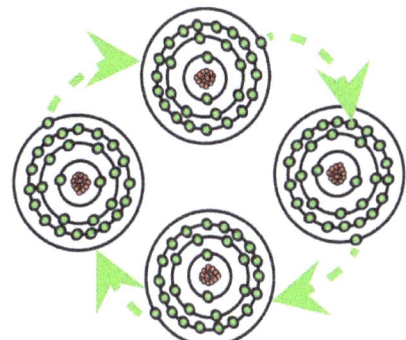

For electrons to move from one atom to another, the conductor must be connected in an unbroken loop known as a **circuit**. This means that as one electron leaves it can be replaced by one from behind. If not connected in a circuit, the electrons cannot flow (move), as the last electron in the conductor has nowhere to go. If the circuit is broken it is said to have lost **continuity**.

Figure 2.22 Atoms in a circle to show how electrons will move from one atom to the next

Magnets

Electricity and magnetism are very closely linked. Both electricity and magnets have positive and negative, or North and South, poles. Both **attract** and **repel**.

- If a copper conductor (wire) is passed by a magnet, the magnetic attraction will move electrons through that copper conductor and create electric current.

- If an electric current is passed through a copper conductor then it will generate an invisible magnetic field.

The magnetic effect of electrical current can be used to make things move (by magnetic attraction or repulsion). That movement can be used to make a motor. The movement of magnets past a conductor can be used to make electric current. This is the principle of a **generator**.

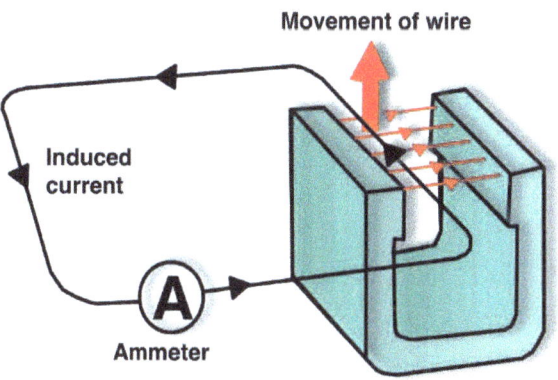

Figure 2.23 The generation of electrical energy by moving a wire through a magnetic field

Mobile Air Conditioning Principles

- Motors turn electrical energy into mechanical energy.
- Generators turn mechanical energy into electrical energy.

> **Circuit** – an unbroken loop.
>
> **Continuity** – refers to an electrical conductor (something that allows electricity to move easily) which is unbroken or complete (i.e. continuous).
>
> **Attraction** – bringing together.
>
> **Repulsion** – pushing away.
>
> **Generator** – a mechanical component that makes electricity.
>
> **Charged** – the storage of electricity, in a battery for example.

Heat

As electric current flows, some of the energy can be turned into heat.

Chemical reaction

Electrical energy can be produced by or converted into chemical energy. Because of this it is possible to store electricity and take it with you in the form of a battery. If you keep a battery **charged**, it provides a portable source of electricity that can be used when needed.

Electrical units

The four main electrical units that you will be using are:

- **Volts**
- **Amps**
- **Ohms**
- **Watts**

They are each named after the person who first described their function.

Volt

Voltage was named after Alessandro Volta. He was the first person to produce moving electricity. Voltage is used to describe the force or pressure in any part of an electrical circuit.

There are two main types of electrical voltage:

- The stored pressure, when everything is switched off.
- The system pressure, when the circuit is switched on.

The stored pressure is known as **EMF** or **electromotive force**.

The pressure found in the circuit when it is switched on is known as **potential difference (Pd)**.

Mobile Air Conditioning Principles

Amp

Amps are the quantity of electricity. The amp was named after a Frenchman, André-Marie Ampère. Amps are used to describe the amount of electric current (moving electricity) in any part of an electrical system.

Ohm

An ohm is the **resistance** to electrical flow. Ohms are named after a German mathematician called Georg Ohm.

Watt

The watt is a measurement of electrical **power** made or used. It is named after a Scottish engineer called James Watts.

> **Volts** – unit of electrical pressure.
>
> **Electromotive force (EMF)** – measurement of electrical voltage with nothing switched on.
>
> **Potential difference (Pd)** – the difference in electrical voltage when a circuit is switched on and energy is being used.
>
> **Amp** – unit of electric current which describes quantity.
>
> **Ohm** – unit of electrical resistance.
>
> **Resistance** – something that slows down movement.
>
> **Watt** – unit of electrical power.
>
> **Power** – the rate at which work is done.

Ohm's law

If any one of the units within a circuit (volts, amps, ohms or watts) is changed (i.e. increased or decreased), this will affect all the other units. Ohms law states:

The current flowing in a circuit is directly proportional to the voltage applied and indirectly proportional to the resistance.

In other words, current flow goes up if voltage is increased and goes down if resistance is increased.

This was explained by Georg Ohm with the following mathematical calculations:

amps = volts ÷ resistance
resistance = volts ÷ amps
volts = amps × resistance

With Ohm's law, if you know two of the electrical measurements, you can calculate the third.

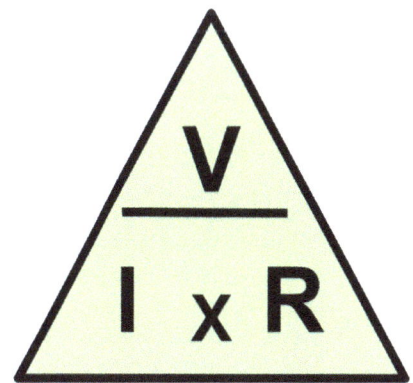

Figure 2.24 Ohms law triangle

Mobile Air Conditioning Principles

The Ohm's law triangle is a good method for calculating the missing unit. It is laid out as shown in Figure 2.24.

In Figure 2.24:

- V = volts (this is sometimes shown as the letter 'E' to represent EMF, but still means volts).
- I = amps (the letter 'I' is used to represent instantaneous current flow).
- R = ohms (the letter 'R' is used for resistance because an 'O' could be confused for a zero).

How to use the triangle
Cover up the unknown unit with your thumb and you are left with the calculation required. For example, amperage is unknown, so cover the 'I' and you are left with V ÷ R (i.e. volts divided by resistance).

Using Ohm's law to help diagnose faults

The relationship between voltage, resistance and amperage can help you to diagnose faults within an electrical circuit. If you take measurements using the different electrical units and then compare them using the Ohm's law calculation, you will be able to work out if the fault is occurring because of:

- **Pressure (volts)**
- If this is lower than expected, component performance is reduced.
- If this is higher than expected, component damage can occur.

- **Quantity (amps)**
- If this is lower than expected, component operation will normally be incorrect.
- If this is higher than expected, component/system operation is being overworked.

- **Resistance (ohms)**
- If this is lower than expected, current may be taking an alternative path to earth (short circuit).
- If this is higher than expected, it will consume electrical energy and reduce system performance.

The power triangle

Watts or power can be calculated in a similar way as:

amps = watts ÷ volts
volts = watts ÷ amps
watts = amps × volts

A power triangle can be used in the same way as Ohm's law. It is laid out as shown in Figure 2.25.

In Figure 2.25:

- W = power (in watts – this is sometimes shown as the letter 'P' to represent power, but still means watts).
- V = volts (this is sometimes shown as the letter 'E' to represent EMF, but still means volts).
- I = amps (the letter 'I' is used to represent instantaneous current flow).

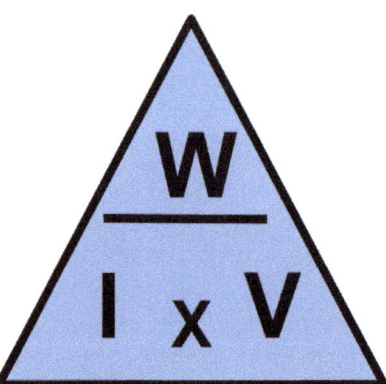

Figure 2.25 Power law triangle

How to use the triangle
Cover up the unknown unit with your thumb, and you are left with the calculation required. For example, amperage is unknown, so cover the 'I' and you are left with W ÷ V (i.e. watts divided by volts).

Mobile Air Conditioning Principles

Series and parallel circuits

Two main types of electrical circuit are used in the construction of motor vehicles:

- Series
- Parallel

Series circuit

In a **series circuit** the consumers are connected in a line one after another. Because they are all in the same circuit, they share the electricity provided depending on the amount of power that they use. If more than one **consumer** is fitted it will only get part of the voltage available.

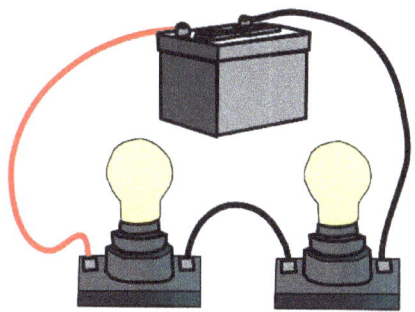

Figure 2.26 A simple series circuit

Series circuit – a circuit with electrical consumers connected in a line, one after another.

Consumer - an item or component that uses up electrical energy, i.e. bulbs, motors etc.

Parallel circuit – a circuit where electrical consumers are connected side by side.

In a series circuit, if consumers are added, total circuit resistance rises.

In a series circuit there is only one path from the power source through all of the components and back to the source. All the consumers share the electricity, so if more than one consumer is fitted it will only get part of the voltage. Voltage drop will occur across each consumer in the circuit, until all available voltage has been used up. This means that if you connect a voltmeter to a series circuit, you will see the reading on the display fall lower after each consumer. This will continue until you see 0 volts after the last component.

If any one of the consumers fail, the circuit is broken and no electricity can flow. The rest of the consumers stop working. This makes series circuits unsuitable for many systems on cars. For example, if you wired a lighting circuit in series, not only would the bulbs glow dimly, but if one bulb broke all of the others would go out.

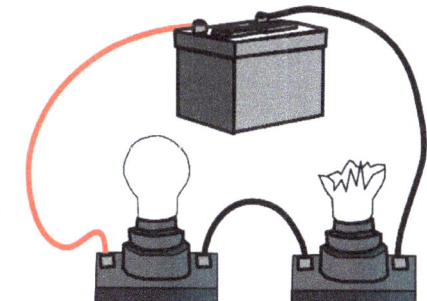

Figure 2.27 A damaged series circuit

With a series system, every time a consumer is added, current draw on the power supply will fall.

Parallel circuit

A **parallel circuit** is one where the consumers are connected next to each other. In a parallel circuit there are multiple parallel paths for the electricity to flow through. Each consumer has its own power supply and earth return back to the power source.

In a parallel circuit, if consumers are added, total circuit resistance falls.

Figure 2.28 A simple parallel circuit

Because each consumer has its own power supply and earth, all the consumers receive the full voltage available and work at full power. So if one consumer in a parallel circuit fails, the others keep working. For example, in a headlight circuit each bulb has its own 12-volt supply and earth return to the battery. If one bulb breaks, the other bulbs will keep working.

Voltage drop also occurs across the consumers of a parallel circuit. However, unlike in a series circuit, if you connect a voltmeter to a parallel circuit, you will see full supply voltage before each component and 0 volts after it.

With a parallel system, every time a consumer is added, current draw on the power supply will increase.

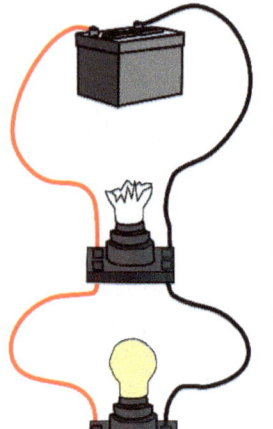

Figure 2.29 A damaged parallel circuit

Cables

Insulated copper wiring is used to transport electricity around the vehicle to where it is needed. Thin strands of copper (a conductor) are bundled together and coated with a plastic shield (an insulator) to help prevent electricity conducting to any other metal components. If this happened, it would cause a short circuit.

Because a large number of wires are used in vehicle construction, the external plastic coating is usually colour-coded (see Figure 2.30). When diagnosing an air conditioning electrical circuit fault you can use these colours to help trace cable routing or identify them on a wiring diagram.

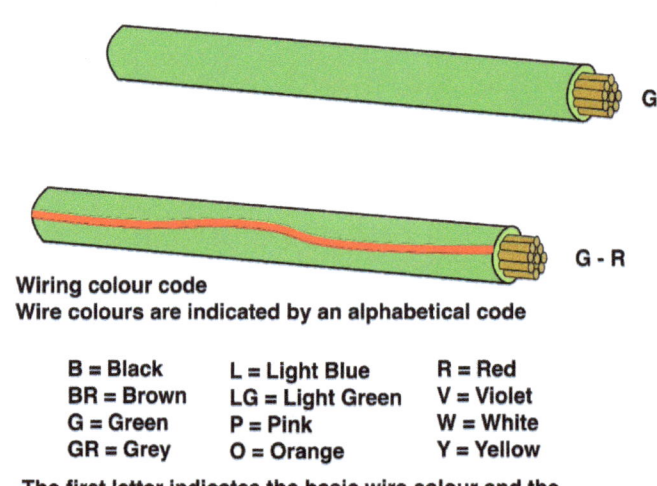

Wiring colour code
Wire colours are indicated by an alphabetical code

B = Black	L = Light Blue	R = Red
BR = Brown	LG = Light Green	V = Violet
G = Green	P = Pink	W = White
GR = Grey	O = Orange	Y = Yellow

The first letter indicates the basic wire colour and the second letter indicates the colour of the stripe.

Figure 2.30 Electric wire with common electrical colour codes

Electrical wires come in different sizes. Because the copper strands are bundled together if one or more strands are damaged electricity can still flow. Automotive wires are normally labelled with the number of strands they contain and the diameter of each strand in millimetres. This gives an indication of the amount of current the wire is able to carry. The thicker the wire, the more electricity it can carry and the less internal resistance it has. The longer the wire, the higher the resistance.

Mobile Air Conditioning Principles

Some typical wire size designations and uses are shown in table 2.6.

Table 2.6 Wire sizes and uses

Number of strands / Wire diameter	Continuous current rating	Uses of the wire
9 / 0.30mm	5.75 amps	Side lamps, tail lamps, reversing lamps, horns, air conditioning control wires
14 / 0.30mm	8.75 amps	Side lamps, tail lamps, reversing lamps, horns, general wiring, air conditioning compressor and circuit wiring
28 / 0.30mm	17.5 amps	Headlamps, fog/driving lamps, windscreen wiper motor, air conditioning compressor and circuit wiring
44 / 0.30mm	27.5 amps	Ventilation fan wiring
65 / 0.30mm	35 amps	Charging cable
84 / 0.30mm	42 amps	Charging cable
97 / 0.30mm	50 amps	Heavy duty charging cable
120 / 0.30mm	60 amps	Heavy duty charging cable
80 / 0.40mm	70 amps	Heavy duty charging cable
37 / 0.71mm	105 amps	Starter/Battery cable
37 / 0.90mm	170 amps	Starter/Battery cable
61 / 0.90mm	300 amps	Starter/Battery cable

The thicker the wire, the more electricity it can carry and the less internal resistance it has. This means that more voltage and current will be available for the component.
The longer the wire, the higher the resistance. This means that less voltage and current will be available by the time it reaches the component.
- Double the length of the wire and you double the resistance.
- Double the diameter and you halve the resistance.

Terminals, connectors and continuity

When a manufacturer designs a car, the electrical wiring used to create the circuits can be bundled together as insulated sections called wiring looms. When the car is assembled, these looms can be routed around the vehicle in the most efficient way, and hidden from view behind panel work, carpets and trims. The wiring looms are made in sections and joined together by **connectors**. At the ends of the looms, **terminals** are used to connect the wires to electrical components.
For electricity to operate the components correctly, the circuits must be continuous and unbroken – this is called **electrical continuity**.

Connector – a component that joins two parts of a circuit together.

Terminal – where the circuit ends (terminates).

Electrical continuity – when an electrical circuit conducts current easily and is unbroken (i.e. continuous), it has electrical continuity.

Mobile Air Conditioning Principles

Earth return systems

Vehicle designers and manufacturers try to keep the amount of wiring used to a minimum. This will save on materials, improve efficiency and reduce costs. Because many vehicles are manufactured mainly from metals (which are good conductors of electricity), it is not always necessary to complete an electrical circuit back to the battery using wire alone.

- The negative end of an electrical circuit wiring can be connected to the vehicle body or chassis. This is called an earthing point.
- The negative terminal of the battery can also be connected to the vehicle body or chassis to complete the circuit. This is called **earth return**.

Earth return systems will only be used for the low voltage (12 volt) systems on a car. If the car is an all-electric or hybrid drive, the high voltage system will run a fully insulated return to the high voltage battery; this includes systems that operate electrically powered air conditioning compressors.

In an earth return system, because the metal of the vehicle body now forms part of the car's electrical system, it is possible to cause damage or injury by mistakenly allowing the positive side of the electrical circuit to come into contact with this earth return. The most common time for this to happen is when you are disconnecting or reconnecting the battery.
To reduce the possibility of injury or damage, you must follow the procedure described below when connecting or disconnecting the cars low voltage battery.

Disconnecting and connecting the battery

When you connect or disconnect the low voltage battery from a vehicle you should remove and refit the terminals in a certain order, as this will help reduce the possibility of a short circuit.

Order for disconnecting: Always remove the negative lead first when you are disconnecting the battery. Once you have disconnected the negative terminal, the vehicle's electrical system is now **open circuit**. Therefore, if the spanner you are using accidentally touches the car's bodywork, both the spanner and the bodywork have the same electrical potential or pressure. If the pressure in an electric circuit is equal, no current can flow (because a difference in pressure creates flow).

Order for connecting: When reconnecting a battery, always connect the positive terminal first and the negative terminal last, for the same reasons.

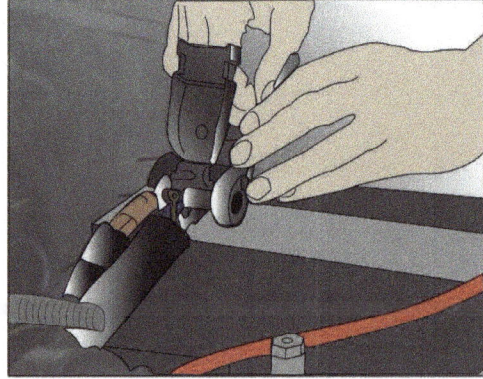

Figure 2.31 A battery terminal being disconnected

Mobile Air Conditioning Principles

> **Earth** – an electrical ground connection, designed to complete a circuit.
>
> **Earth return** – using the metal chassis to complete the car's electrical circuit.
>
> **Open circuit** – a broken electrical circuit where no electricity can flow.

Electric symbols and wiring diagrams

Many electric circuits are designed using wiring diagrams. To help with the design and reading of wiring diagrams, manufacturers use symbols to represent electrical system components.

Common electrical and electronic symbols

Examples of some common electrical and electronic symbols are shown in table 2.7.

Table 2.7 Common electrical and electronic symbols

Symbol	Definition
Switch symbols 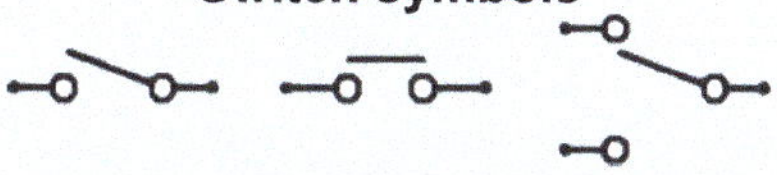	Switch: a component that controls electrical current in a circuit.
Battery symbol 	Battery: source of portable electrical power.
Electric motor symbol 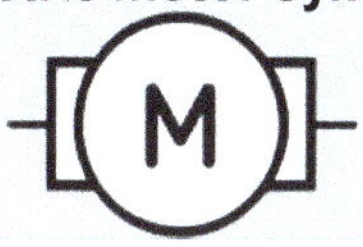	Motor: a component that uses electromagnetism to provide mechanical movement.
Cables crossed but not joined 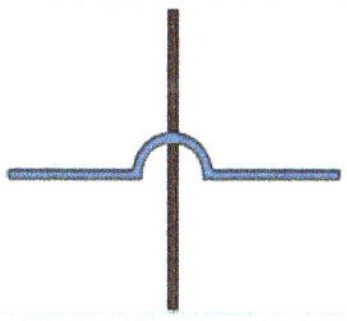	Cables crossed: electrical wiring that is crossed over on a diagram, but not joined.
Fuse symbol 	Fuse: a component that protects an electric circuit from excess current.
Cables crossed and joined 	Cables joined: electrical wiring on a diagram that is crossed over, but also joined.
Bulb symbol 	Bulb: a component that converts electrical heat into light.
Earth/Ground symbol 	Earth: an electrical ground connection, designed to complete a circuit.

Mobile Air Conditioning Principles

Table 2.7 Common electrical and electronic symbols

Symbol	Definition
Transistor symbols	Transistor: An electronic component that acts as a switch with no moving parts.
Resistor symbol	Resistor: a component that slows down and restricts the flow of electricity.
Electric relay symbol	Relay: A remote, electromagnetic switch.
Variable resistor symbol	Variable resistor: a resistor that can be adjusted to control the flow of electricity.
Electrical diode symbol	Diode: an electronic component that acts as a one-way valve for electricity.

Wiring diagrams

A simple wiring diagram of an air conditioning systems electrical circuit is shown in Figure 2.32.

Study this to help you get used to the layouts and symbols.

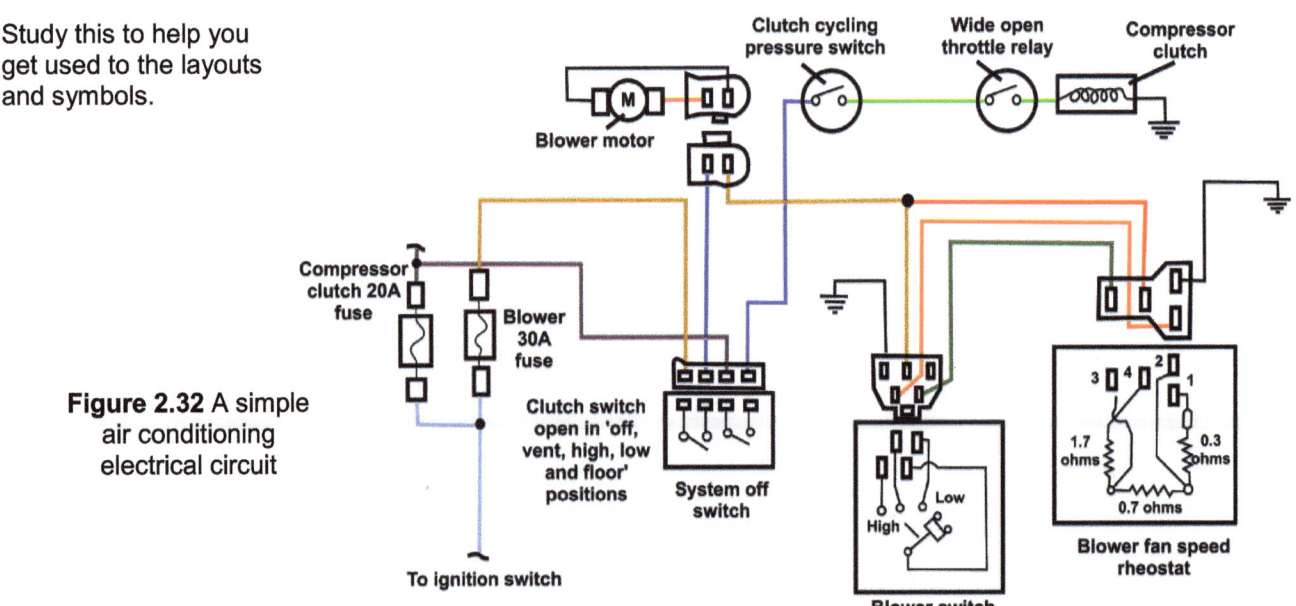

Figure 2.32 A simple air conditioning electrical circuit

Mobile Air Conditioning Principles

Switches

A switch is an electrical component that is designed to control the circuit. This can be as simple as making or breaking the electrical circuits so that current can be turned on and off, or the switch could be a variable resistor (potentiometer) that can regulate the flow of electricity (as with a dimmer switch). Switches give variable control to air conditioning circuits, allowing the user to select various operating conditions.

Circuit relays

A **relay** is an electromagnetic switch. It is designed to allow a small current to switch a much larger current. A coil of wire is wrapped around a soft iron core. When a small electric current is passed through it, it creates an **electromagnet**. This electromagnet is then able to open or close a switch inside the relay unit.

Relay – an electromagnetic switch used to control large currents.

Electromagnet – a metal core made into a magnet by passing electric current through a surrounding coil.

Uses of relays

Relays can be useful in circuits that use large amounts of electric current.

In an air conditioning control circuit, a relay can be used to allow smaller lightweight switches to be used on the dashboard of the car, while still being able to operate high current consuming devices such as the condenser cooling fan.

If relays weren't used in electric circuits:

• Large-diameter wires would have to be routed all over the car, increasing cost and overall weight.

• Any electrical switches would be under excessive strain and likely to fail early.

• The length of wire used in the circuit would increase overall resistance. This would reduce the amount of electricity available by the time it reached the consumer, reducing its performance.

When relays are used in an electric circuit:

• Wiring carrying the heavy-duty current can be designed to take the shortest route to the consumer (reducing overall circuit resistance) but will be controlled by the mechanical switch part of the relay, mounted nearby.

• A lightweight switch can now be used by the driver to control the electromagnet (which only consumes a small amount of current) within the relay.

• When the driver operates the switch to control the circuit, a magnetic field is created inside the relay, causing the heavy duty switch to open or close.

Types of relay

There are three main types of relay:

- M4 – This type of relay has four electrical terminals. When operated, it switches on a circuit (makes the circuit).

- B4 – This type of relay has four electrical terminals. When operated, it switches off a circuit (breaks the circuit).

- Double throw – This type of relay normally has a five electrical terminals. When operated, it switches between two electrical circuits.

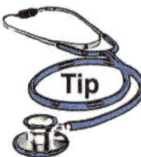

When testing a relay, don't disconnect it, otherwise you're breaking the circuit and stopping it from working anyway.

The DIN terminal numbers for relay connections can be found in the appendix.

Current-consuming devices

The scientific law of conservation states:

'Energy cannot be created or destroyed but only released and converted into some other form of energy.'

If electrical energy in a circuit is not used up, it will be changed into heat and a fire could result.

Something is needed in an electrical circuit to do some useful work, and this is called a consumer. Some examples of consumers used in air conditioning systems are:

- Ventilation fans
- Condenser cooling fans
- Ventilation flap servo motors
- Compressor clutch
- Dashboard control illumination and displays

Electrical consumers are designed to work when the system voltage is switched on. If working correctly, these consumers should use all of the electrical pressure available. For example, if 12V are supplied to the consumer, there should be 0V after the consumer. This is called **volt drop** and is a key element used when diagnosing electrical faults, see Chapter 3.

Volt drop – the fall in electrical voltage when current flows through a consumer.

Mobile Air Conditioning Principles

Motors and fans

A simple motor can be made by passing an electric current through a coiled wire that is wound around a central shaft called the **armature** – this creates an electromagnet.

• The electric current produces an invisible magnetic field, which is repelled (pushed away) or attracted (pulled towards) by the permanent magnets surrounding it. This causes the armature to turn.

• Once the armature has turned out of the magnetic field, it would normally stop.

• To keep it rotating, the polarity of the electricity passing through the electromagnet mounted on the armature must be changed. This is done by a component called a **commutator**.

• Two spring-loaded electrical contacts called **brushes** are mounted on the end of the armature to maintain electrical connection with the commutator as the shaft rotates.

• When electric current is switched off, the motor will stop.

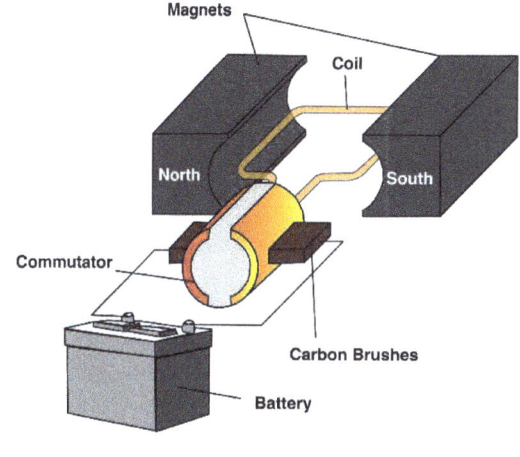

Figure 2.33 A simple electric motor

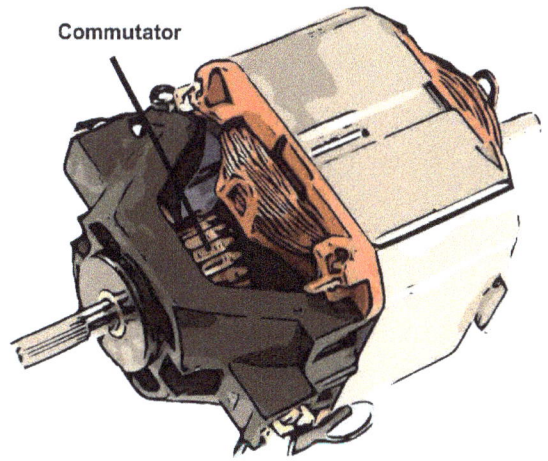

Figure 2.34 Commutator

Armature – the central shaft of an electric motor.

Commutator – a segmented electrical contact mounted on the end of a motor armature, designed to change electrical polarity as the motor turns.

Brushes – spring-loaded electrical contacts that transfer current to the rotating armature.

The end of the armature shaft on a simple electric motor can be attached to:

• A set of fan blades, used to direct air through the condenser to help with cooling.

• A ventilation fan used to control the speed of the air passing through the air conditioner and passenger compartment.

Mobile Air Conditioning Principles

Fuses

If you allow electrical current to flow in a circuit without passing through a consumer, then the energy will be converted into heat. This can rapidly cause the circuit wiring and components to burn out, which will be expensive to repair.

To protect the circuit, fuses or circuit breakers are commonly used. A fuse is a weak link placed in the circuit. It is a thin piece of wire (with a current rating just above that of the current intended to flow in the system) that is fitted in line (series). (See Chapter 3 for common fuse colours and amperage ratings).

This relatively inexpensive component is designed to burn out if a rapid increase in current flow occurs to prevent any further damage to the rest of the circuit. Once the fuse has burnt out, an open circuit exists and no further current can flow (electricity stops), hopefully saving more expensive parts. Fuses come in different sizes, shapes and types, including glass, ceramic and blade. Blade fuses are the most common type found on light vehicles today.

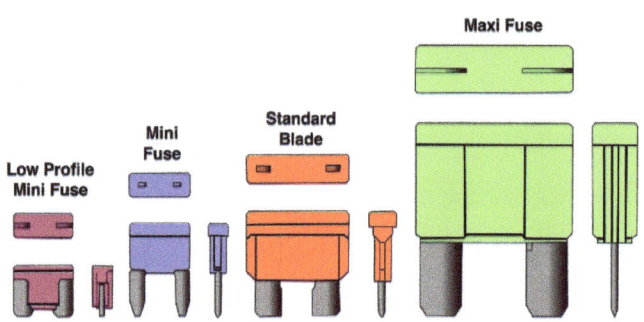

Figure 2.35 Blade fuses

Sensors

To help maintain control of air conditioning and climate control systems, electrical/electronic sensors are often used. These sensors are placed at various points on an air conditioning system or around the car and they pick up on how the system is operating or needs to operate. They send signals to an electronic control unit ECU, which processes the information and operate actuators. For more information on sensors and actuators used in air conditioning and climate control systems see Chapter 3.

Electronic thermostats

A thermostat is a device used to control system temperatures. With an electronic thermostat a small electrical heating element is used to control the opening and closing of a valve which can regulate the flow of refrigerant in an air conditioning system. An ECU is able to control the current flow to the heating element warming wax or a liquid which expands and opens the valve in the thermostat.

Electrical power sources

The battery

Because vehicles are mobile, the electrical energy source must be portable. The electrical energy needed for a light vehicle can be carried in a chemical container called a battery.

How the battery works

A standard low voltage lead-acid battery, the type often found in cars, contains a number of sections called cells (see Figure 2.36). Each cell is capable of producing approximately 2.1V. A standard battery contains six cells linked together, creating a battery with a voltage of 12.6V. This is rounded down to 12, so we say that the battery has a voltage of 12V.

Mobile Air Conditioning Principles

Each cell contains a number of lead plates, which are chemically different:

• The negative plate is made of lead.

• The positive plate is made of lead peroxide.

To prevent the plates from touching each other and causing a **short circuit**, thin sheets of material called separators are inserted between them.

Lead peroxide contains extra oxygen compared with normal lead. This means that the positive plate is chemically different from the negative plate, and it would like to share its electrons with the negative plate. If connected in a circuit, the electrons are allowed to move from the positive plate to the negative plate. This creates electric current, which provides the energy to power components in the vehicle.

The first half of the circuit is made with an **electrolyte** (a liquid that the cell is filled with), which consists of sulphuric acid and **deionised** water. The electrolyte covers the plates and allows electrons to move from one plate to another. The top of each plate is then connected to the rest of the circuit. The circuit must contain a consumer to use up the electrical potential (energy).

When the circuit is complete, the electrons combine with the electrolyte and move from one plate to another as a chemical reaction, creating current.

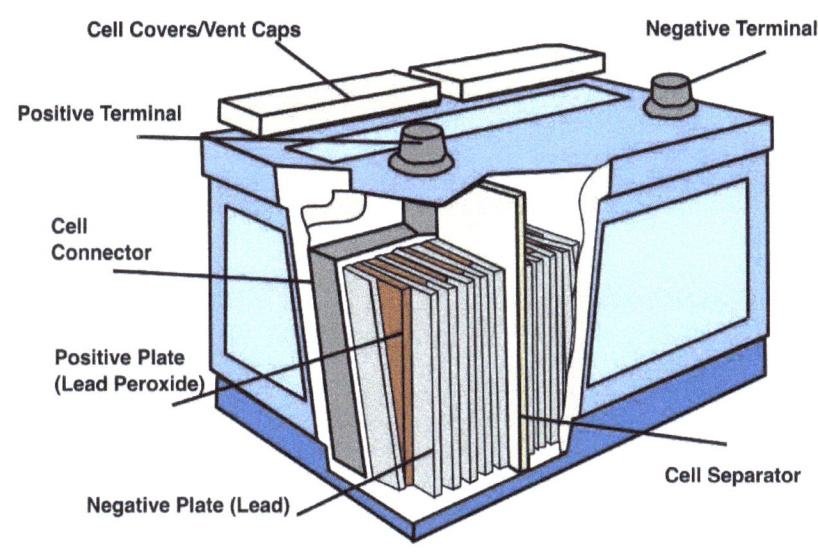

Figure 2.36 Components of a vehicle's lead-acid battery

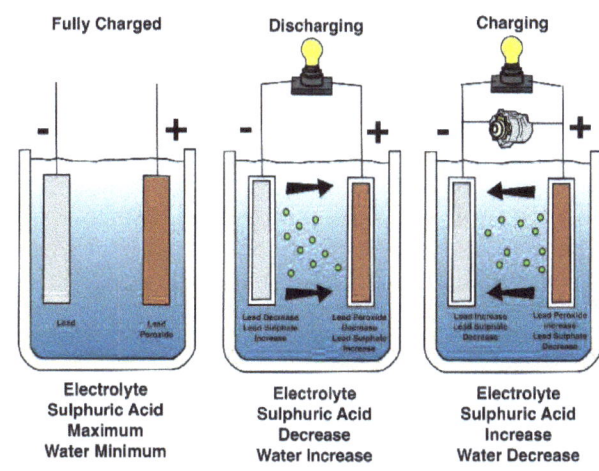

Figure 2.37 The chemical reactions taking place inside a lead-acid battery

Short circuit – electricity taking a short cut and not travelling the full length of the circuit.

Electrolyte – the liquid inside a battery cell, usually made from deionised water and sulphuric acid.

Deionised water – water that has been treated to remove any electrical charge.

Mobile Air Conditioning Principles

Check your knowledge

1. What is the purpose of the compressor within the Air Conditioning system?
 a To raise the pressure and raise the temperature of the refrigerant.
 b To lower the pressure and raise the temperature of the refrigerant.
 c To lower the pressure and lower the temperature of the refrigerant.
 d To raise the pressure and lower the temperature of the refrigerant.

2. Oxygen Free Nitrogen OFN is used to:
a Remove R12 refrigerant when converting the system to R134a refrigerant (retro-fit).
b Flush air conditioning systems.
c Help locate air conditioning system leaks.
d Remove moisture from air conditioning systems.

3. A receiver drier is fitted to an air conditioning system:
a with a fixed orifice tube.
b with an AC system fitted a thermal expansion valve TXV or a fixed orifice tube FOT.
c only on an AC system containing R12 refrigerant.
d with a thermal expansion valve TXV.

4. If the air conditioning system has been contaminated, what action should the MAC technician take?
a None of the answers.
b Recharge the air conditioning system.
c Replace the faulty components and recharge the air conditioning system.
d Replace any faulty components and ensure that the air conditioning system is flushed to remove all of the contaminants before recharging the system.

5. Which of the following units represents electrical pressure?
a Volts
b Amps
c Ohms
d Watts

6. Which piece of equipment can be used to help check if an air conditioning system is contaminated with hydrocarbons?
a A thermometer.
b A refrigerant identifier.
c A manifold gauge set.
d A set of weighing scales.

Mobile Air Conditioning Principles

7. Which of the following is not an air conditioning system electrical current consumer?
a Fuse.
b Ventilation fan.
c Compressor clutch.
d Flap control servo motor.

8. An electromagnetic compressor clutch with a resistance of 6 ohms being powered by a 12 volt battery will use how much current?
a 2 watts
b 2 volts
c 2 ohms
d 2 amps

9. What is the purpose of the Air Conditioning system condenser?
a To change the hot refrigerant vapour into a liquid.
b To change the hot refrigerant liquid into a vapour.
c To change the hot refrigerant liquid into a cool vapour.
d To change the cool refrigerant liquid into a vapour.

10. What is the purpose of the pollen filter?
a To remove unwanted particles before it enters the interior of the vehicle.
b To prevent damage to the evaporator.
c Prevent unpleasant odours from the heating and ventilation system.
d To reduce the humidity of the air entering the interior of the vehicle.

Answers: 1a, 2c, 3d, 4d, 5a, 6b, 7a, 8d, 9a, 10a

Diagnostics for Mobile Air Conditioning & Climate Control

Chapter 3 Diagnostics for Mobile Air Conditioning and Climate Control

This chapter will help you develop knowledge and understanding of air conditioning and climate control. This will enable you to conduct effective diagnosis and repairs of system faults. It supports you by providing knowledge that will help you when undertaking both theory and practical assessments. Remember to work safely at all times and observe the relevant environmental, health and safety regulations, while developing air conditioning and climate control diagnostic routines that are systematic and effective.

Contents
Component identification, function & operation for AC & climate control 90

Diagnosis of air conditioning system faults 108

Diagnosis of climate control system faults 112

How to prepare for assessment 143

Safe working when handling refrigerant

There are many hazards associated with the handling of refrigerant used in air conditioning systems. You should always assess the risks involved with any maintenance or repair routine before you begin and put safety measures in place.
You need to give special consideration to the possibility of:
• Electric shock caused by incorrect circuit connection
• Injury caused by being caught up in rotating system components
You should always use appropriate personal protective equipment (PPE) when you work on these systems. Make sure that your selection of PPE will protect you from these hazards.

Personal Protective Equipment (PPE)

Table 3.1 PPE required when working on vehicle air conditioning systems

PPE	Recommendations
Overalls	Overalls provide protection from coming into contact with oils and chemicals.
Gloves	Fluroelastomer gloves provide protection from fluorinated refrigerants and help protect the hands from frostbite.

Diagnostics for Mobile Air Conditioning & Climate Control

Table 3.1 PPE required when working on vehicle air conditioning systems

PPE	Recommendations
Protective footwear 	Safety boots protect the feet from a crush injury and often have oil and chemical resistant soles. Safety boots should have a steel toe-cap and steel mid-sole.
Goggles 	Safety goggles reduce the risk of small objects or refrigerants coming into contact with the eyes.
Bump cap/Hard hat 	A bump cap or hard hat protects the head from bump injuries when working under cars.

Vehicle Protective Equipment (VPE)

To reduce the possibility of damage to the car, always use the appropriate vehicle protection equipment (VPE):

 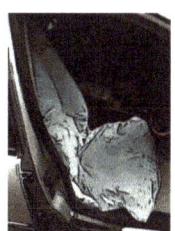  

Wing covers   Seat covers   Steering wheel covers   Floor mats

Information sources

The complex nature of air conditioning and climate control systems requires you to have a good source of technical information and data. In order to conduct maintenance and repair procedures, you need to gather as much information as possible before you start.

Sources of information may include:

Table 3.2 Possible information sources

Verbal information from the driver	Vehicle identification numbers
Service and repair history	Warranty information
Vehicle handbook	Technical data manuals
Workshop manuals/Wiring diagrams	Safety recall sheets
Manufacturer specific information	Information bulletins
Technical helplines	Advice from other technicians/colleagues
Internet	Parts suppliers/catalogues

Diagnostics for Mobile Air Conditioning & Climate Control

Table 3.2 Possible information sources

Jobcards	Diagnostic trouble codes
Oscilloscope waveforms	On vehicle warning labels/stickers
On vehicle displays	Temperature readings

Always compare the results of any inspection, testing or diagnosis to suitable sources of data. Remember that no matter which information or data source you use, it is important to evaluate how useful and reliable it will be to your safety, maintenance and repair routine.

Air conditioning and climate control

Air conditioning and climate control are both comfort and convenience systems. They operate on a refrigerant cycle which is common to both. The main difference is that air conditioning operation and function are manually set by the driver, and climate control uses a number of sensors to inform an ECU which then automatically controls the function and operation of the system.

The component parts of air conditioning and climate control systems

Compressor

So that the refrigerant cycle can operate correctly, there needs to be a change of state between a gas to a liquid and a liquid back into a gas. The compressor is an engine or electrically driven pump mechanism, designed to raise the pressure of a refrigerant gas. The pressure rise is needed in order to raise the boiling point of the refrigerant. This way, a refrigerant with a very low boiling point, such as Tetrafluoroethane at -26.3°C can be cooled sufficiently in the condenser for it to become a liquid. The compressor is often mounted on the vehicles internal combustion engine and driven by an auxiliary belt connected to the crankshaft pulley. The drive belt may be a 'V' type which fits a pulley profile known as an 'A' groove or a multi-rib type which fits a pulley known as a 'polyvee'. The mounting of the compressor is adjustable to allow the drive belt to be tensioned during maintenance and repair procedures. Because compressors are mechanically operated they require lubrication. For a description of the lubrication oils used in compressors, see Chapter 2.

If the vehicle is a hybrid drive or all electric drive, operating the compressor from an internal combustion engine is not always an option; on a hybrid vehicle the internal combustion engine is not always running and an electric vehicle has no engine. In these designs, the compressor is often powered from a high voltage electric motor.

Figure 3.1 A high voltage electric powered compressor

A number of different types of compressor are available.

Piston compressors

A piston type compressor operates in a similar manner to a standard reciprocating internal combustion engine, but in reverse, see figure 3.2. An engine driven pulley, rotates a crankshaft driving a piston or pistons up and down inside a cylinder bore. Mounted above the piston are a pair of one-way valves; one inlet and one outlet. These

Diagnostics for Mobile Air Conditioning & Climate Control

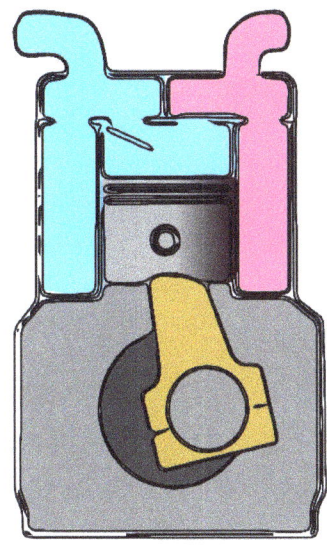

valves are often passive, meaning that they are operated by a difference in pressure alone, with no driving mechanism. As the piston is moved up and down by the operation of the crankshaft, it creates a pumping action. On the down stroke, the difference in pressure above the inlet valve and the depression caused by the descending piston is enough to open the valve against its spring, drawing in refrigerant gas. As the piston moves upwards the inlet valve closes, sealing the pumping chamber and raising the pressure of the gas. Eventually, the rising pressure in the pumping chamber is enough to overcome the spring force of the outlet valve. The refrigerant gas is forced out of the valve, raising the pressure of the first half of the air conditioning system.

Figure 3.2 A piston type compressor

Vane compressors

A vane type compressor uses a set of small paddle wheels attached to a rotating central shaft, see Figure 3.3. The rotating shaft is mounted in an offset position inside a circular pumping chamber. As the central shaft is turned by the engine, or electric motor in the case of a hybrid electric vehicle, the paddle wheels are held in contact with the pumping chamber walls forming a gas tight seal. In this design of compressor, as there is no reciprocating action, inlet and outlet valves are not needed. Instead, inlet and outlet ports are used to transfer refrigerant through the compressor from the low pressure side to the high pressure side. As the vane element is turned, it uncovers the inlet port from the low pressure side of the air conditioning circuit. Because the central shaft is offset, this creates an expanding chamber, causing a depression which draws in the gaseous refrigerant. As it continues to rotate, the vanes close off the inlet port and force the gas into a smaller and smaller space, creating compression which raises the pressure and temperature of the refrigerant. Eventually the vanes uncover the outlet port, and the high pressure refrigerant gas is forced out of the compressor towards the condenser. As the shaft continues to turn, the whole process is repeated.

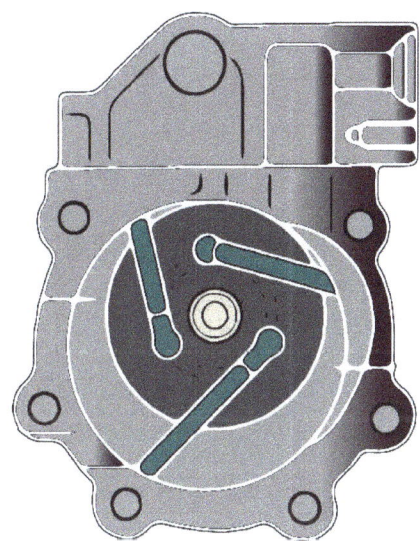

Figure 3.3 A vane type compressor

Swash plate compressors

A Swash plate compressor is a form of reciprocating piston compressor, but instead of the pistons moving up and down attached to a crankshaft, they are often small plungers that are pushed longitudinally, backwards and forwards along the length of the compressor, see Figure 3.4. The main driving element for these plungers is a component known as a Swash plate. The Swash plate is a metal disc set at an angle in the centre of a rotating shaft. As the shaft is turned by the engine, the Swash plate is also rotated, causing it to move backwards and forwards with a wobbling action. This wobbling action can be used to drive the plungers/pistons inside the compressor creating the pumping action. Mounted at one end of the plunger/piston are a pair of one-way valves, one inlet and one outlet. These valves are often passive, meaning that they are operated by a difference in pressure alone, with no driving mechanism. On the back stroke, the difference in pressure behind an inlet valve and the depression caused by the retracting plunger/piston is enough to open the valve against its spring, drawing in refrigerant gas. As the plunger/piston moves forwards the inlet valve closes, sealing the pumping chamber and raising the pressure of the gas. Eventually, the rising pressure in the pumping chamber is enough to overcome the spring force of the outlet valve. The refrigerant gas is forced out of the valve, raising the pressure of the first half of the air conditioning system.

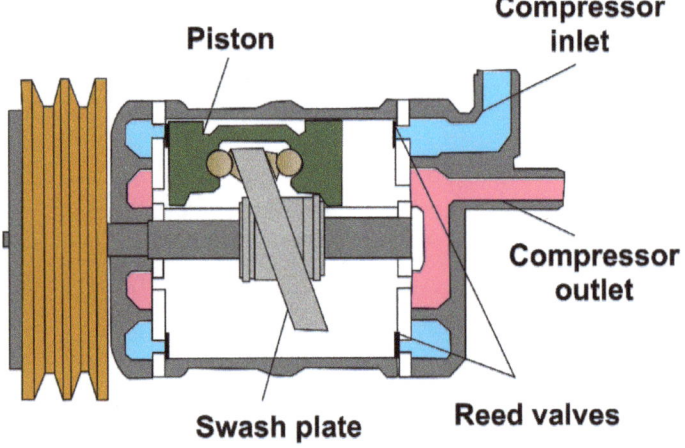

Figure 3.4 A Swash plate compressor

The Swash plate inside a compressor can have a sprung-loaded mounting on the central shaft. The spring pressure, will hold the Swash plate at an angle to create the wobbling action. If the system gas pressure rises too high, this will cause the plungers to overcome the spring force, causing the Swash plate to become vertical reducing the pumping action and lowering the pressure. Once the pressure falls, the spring force returns the Swash plate to an angle, causing it to wobble once again. This means that this type of compressor can be made self-regulating for pressure.

The Swash plate of some compressors can be attached to an actuator arm which can change the angle of the plate depending on system requirements; this is often done electronically. This design of compressor means that two-stage air conditioning can be made available.

Stage one will create a shallow angle at the Swash plate reducing the pumping and output. This will give less air conditioning efficiency but reduce strain on the vehicles engine, improving performance, fuel economy and reducing exhaust emissions.

Stage two will create a steeper angle at the Swash plate, increasing pumping output. This will give better air conditioning performance but increase strain on the vehicles engine. This will reduce engine performance, fuel economy and increase exhaust emissions.

Diagnostics for Mobile Air Conditioning & Climate Control

Some compressors are able to actuate the angle of the swash plate in accordance with system loads and demands. These are known as variable displacement compressors and when controlled by an ECU can be used with fixed orifice tube systems to help regulate refrigerant pressure and flow.

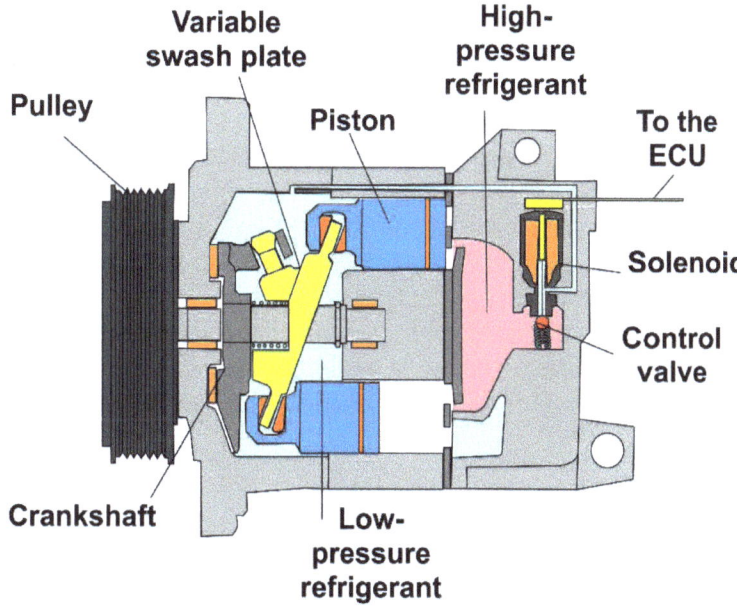

Figure 3.5 A variable displacement compressor

Scotch yoke compressors

A scotch yoke compressor is a type of piston compressor which does not use a traditional type of crankshaft to drive the pistons. Instead a sliding yoke is connected to an offset drive pin on a central shaft driven by the vehicles engine or an electric motor. The action of the scotch yoke will turn the rotational movement of the shaft into a linear movement of the yoke, which can then be used to drive reciprocating pistons, see Figure 3.6. If carefully designed this mechanism can be used to drive four pistons in a very compact and efficient manner. Mounted above each of the pistons are a pair of one-way valves, one inlet and one outlet. These valves are often passive, meaning that they are operated by a difference in pressure alone, with no driving mechanism. As each of the pistons is moved backwards and forwards by the operation of the scotch yoke mechanism, it creates a pumping action. On the back stroke, the difference in pressure behind the inlet valve and the depression caused by the retracting piston is enough to open the valve against its

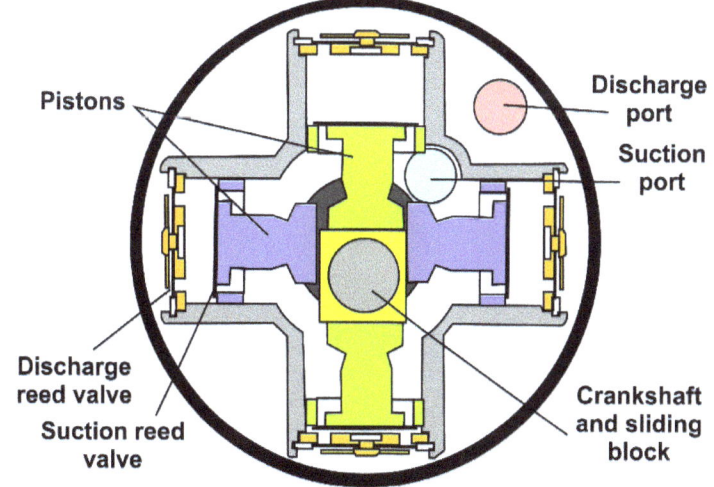

Figure 3.6 A scotch yoke compressor

spring, drawing in refrigerant gas. As the piston moves forwards the inlet valve closes, sealing the pumping chamber and raising the pressure of the gas. Eventually, the rising pressure in the pumping chamber is enough to overcome the spring force of the outlet valve. The refrigerant gas is forced out of the valve, raising the pressure of the first half of the air conditioning system.

Scroll compressors

A scroll compressor is one of the most efficient designs for use in an air conditioning or climate control system, but due to its complex design, can be one of the most expensive to produce. In this type of compressor, a spiral design known as a scroll, is machined into a drive component and the compressor housing which fit together with a very fine **tolerance**, see Figure 3.7. A central shaft is connected to the drive component by a cam mechanism, which when turned by the engine or an electric motor, **oscillates** the machined drive spiral in the housing spiral component. This oscillating action creates a depression at one end of the spiral scroll, causing refrigerant gas to be drawn in through an inlet port. As the drive component continues to oscillate, the inlet port is closed off and the trapped refrigerant is forced around the scroll into an ever decreasing space created by the spiral design; this raises the pressure and temperature of the refrigerant. Eventually the scroll uncovers the outlet port, and the high pressure refrigerant gas is forced out of the compressor towards the condenser. As the scroll continues to oscillate, the whole process is repeated.

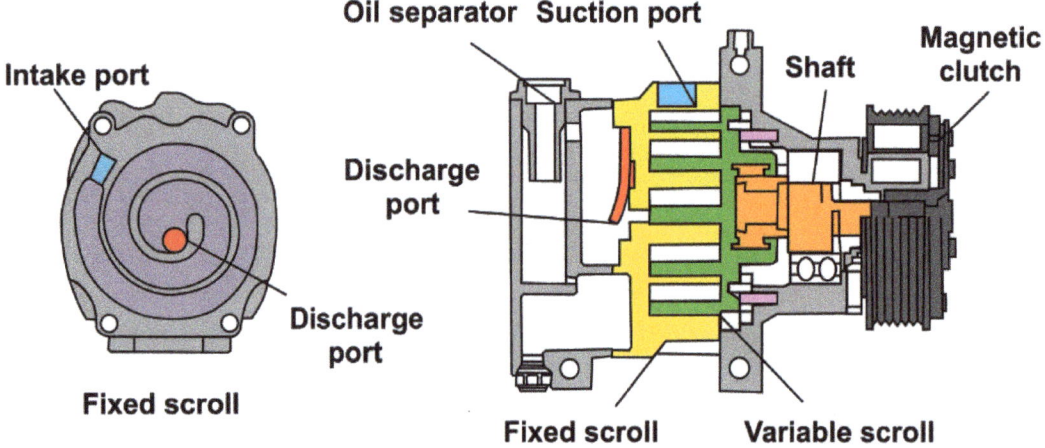

Figure 3.7 A scroll type compressor

Tolerance - an allowable difference between two measurements.

Oscillate - to move backwards and forwards or up and down with a repeating action.

Bimetallic strip - a strip of two different metals joined together, which bends with a rise in temperature.

Diagnostics for Mobile Air Conditioning & Climate Control

Thermal switches

Switches are used to control the flow of electricity by making and breaking an electric circuit. Thermal switches are temperature sensitive components that can be used in an air conditioning system to automatically control the operation of various electrically operated components.

For example, a thermal switch can:

- Switch on a condenser cooling fan if its temperature rises too much.
- Switch off a compressor clutch if the evaporator temperature falls too low.

Most thermal switches will be constructed using **bimetallic strips**. A bimetallic strip is two thin strips of different metal stuck together. As metal is heated, it expands and as metal is cooled, it contracts. If a bimetallic strip is heated or cooled, the two metal strips will expand or contract at different rates causing it to bend. If one end of the bimetallic strip is formed into a switch contact, as the metal bends it can be used to make or break an electric circuit. These bimetallic strips are manufactured into small housings, creating a self-contained unit which can be mounted at strategic positions in the refrigeration circuit to help regulate temperatures through electronic control.

High pressure relief valves

Air conditioning systems are manufactured to have a maximum design operating pressure. If a fault occurs, such as a blockage, pressures could increase past safe levels causing system/component damage or even explosion. The purpose of a high pressure relief valve is to keep the refrigerant circuit within its design operating pressure. The relief valves can be mounted anywhere on the high pressure side of an air conditioning system, but are most commonly found on the compressor or are manufactured as part of a thermal expansion valve (TXV). The relief valve will often consist of a sprung loaded plunger or bearing which is held against a seal, until a maximum pressure is reached. Once this pressure has been reached or exceeded, the plunger or bearing is lifted off its seat against the force of the spring and refrigerant is allowed to leak away into the low pressure side of the circuit.

Superheat and thermal limiters

Correct refrigerant flow plays an important role in maintaining the operation of the air conditioning compressor. The refrigerant carries lubrication oil to the compressor (as a mist in the vapour) and the cold refrigerant returning from the evaporator also provides the cooling for the compressor. When the air conditioning system is low on refrigerant, the evaporator goes into a superheat condition. That means the refrigerant vapour in the evaporator is being overheated. When the system is in a superheat condition, both oil flow and compressor cooling will be compromised. On some early air conditioning systems, to prevent catastrophic compressor failure, a superheat switch was mounted on the rear of some compressors which could activate a thermal limiter. A thermal limiter is a form of circuit breaker designed to protect the air conditioning compressor by stopping the clutch from engaging whenever the superheat switch senses that the system is low on refrigerant.

High pressure, low pressure and trinary switches

To provide system control and protection, many air conditioning circuits use high and low pressure switches, instead of a superheat switch and thermal limiter. These pressure switches are normally mounted in the high pressure circuit and will disengage the compressor clutch when a predetermined system pressure is reached.

The high pressure switch operates if pressures exceeds an upper limit.

Diagnostics for Mobile Air Conditioning & Climate Control

Situations which may cause this include:

- Too much refrigerant (system over-gassed)
- Blockage
- Ice in the system
- Condenser temperature too low

The low pressure switch operates if pressure exceeds a lower limit.

Situations which may cause this include:

- Too little refrigerant (system under-gassed)
- Leakage
- Evaporator temperature too high (superheat)

Figure 3.8 A trinary switch

These two pressure functions may be combined in one unit which senses high and low pressure.

In a system with the correct amount of refrigerant, high pressures can be used as an indication of condenser temperature. A third function of a high/low pressure switch can be to operate the condenser cooling fan as system pressures rise. This three function switch is often called a trinary switch. An example of the operating pressures used by a trinary switch in an R134a system are shown in table 3.3.

Table 3.3 Examples of the operating pressures used by a trinary switch

Trinary switch function	Switch off pressure (psi)	Switch on pressure (psi)
Low pressure switch	29 + or - 4 psi	31 + or - 5 psi
Condenser fan switch	181 + or - 22 psi	239 + or - 17 psi
High pressure switch	435 + or - 29 psi	348 + or - 29 psi

Clutch assembly

The drive pulley of many compressors are attached to the pumping mechanism by an electromagnetic clutch assembly. A clutch is a component that can engage and disengage two rotating components. The compressor pulley of an engine driven unit is mounted on a bearing which allows it to spin freely without turning the central shaft. Either side of the pulley are two clutch driving members, which in the disengaged position do not touch the pulley. The inner clutch driving member is attached to the central shaft and contains a strong electromagnet. When powered, this electromagnet attracts the outer driving member of the clutch mechanism gripping the drive pulley tightly between the clutch members. The pulley is now attached securely to the central shaft which is rotated to provide compressor action. When power is removed from the electromagnet, the clutch mechanism disengages, the pulley once again freewheels and compressor action stops.

Figure 3.9 Compressor clutch

Diagnostics for Mobile Air Conditioning & Climate Control

Notes: The engaging and disengaging of the compressor clutch can be used to control the refrigerant cycle. Cutting the compressor in and out to regulate system operation and temperature is known as 'cycling'.

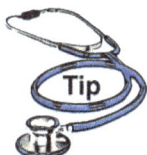

Tip: Because the compressor clutch drive members are metal plates, they can often be heard engaging with a loud 'click'. If a fault occurs in the air conditioning system, such as a lack of refrigerant, the clutch is prevented from engaging due to the low pressure switch. When diagnosing air conditioning faults, listening for the compressor clutch 'click' may help provide an indication that the system is trying to operate normally.

Condensers

A key part of the air conditioning cycle is to remove a large proportion of the heat energy absorbed in the refrigerant. Most of the heat energy in the refrigerant gas has come from the process of compression and the remainder has been absorbed from the air inside the passenger compartment. As heat is an energy and cannot be destroyed it must be efficiently radiated to atmosphere and this is the job of the condenser. Mounted at the front of the car in a position that receives the most air flow, a good condenser should be able to radiate heat from the refrigerant with very little loss in pressure. The design of the condenser will have a large effect on its ability to operate efficiently.

Two main designs are common:

Serpentine - in this design the hot refrigerant gas enters the top of the condenser and into a pipe that snakes backwards and forwards across the width of the radiator in a zigzag pattern. This gives a long period of exposure to the cooling air flow over the condenser core. To assist with the removal of heat, small, thin metal fins are attached to the outside of the pipe which increase the overall surface area. As the refrigerant leaves the bottom of the condenser, enough heat should have been removed to ensure the refrigerant has condensed into a high pressure liquid.

Parallel flow - in this design the hot refrigerant gas enters the condenser at the top of a manifold section at one side of the unit (sometimes called a header tank). In order for the refrigerant to reach the outlet of the condenser it must pass through a series of flat tubes which contain a number of smaller pathways which break the flow up into tiny streams. These tiny parallel pathways are very efficient at dissipating heat.

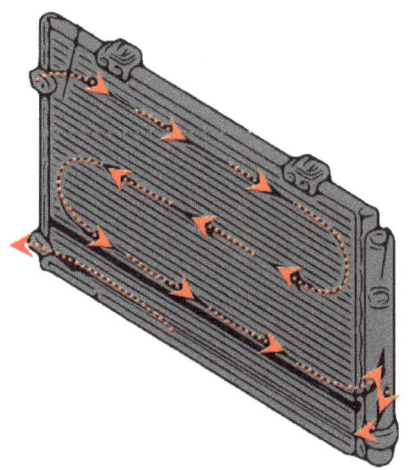

Figure 3.10 A condenser radiator

Diagnostics for Mobile Air Conditioning & Climate Control

To assist with effective cooling, condensers will be fitted with an electric fan which can provide extra air flow when the vehicle is stationary or if the condenser temperature or pressure rises too high. Depending on the design and space available, a single or a twin fan can be fitted.

Figure 3.11 A condenser fan assembly

Due to the design of a parallel flow condenser, it cannot be flushed (see Chapter 2). If the condenser becomes contaminated or blocked with dirt it must be replaced.

Some condenser units are actually two condenser cores mounted one on top of the other. These 'dual-core' condensers will have a combined receiver drier integrated between the two cores, sometimes known as a 'modulator'. The flow of refrigerant through the first condenser will dissipate heat, allowing some of the gas to transform into a liquid state which is temporarily retained in the modulator. The remaining refrigerant gas is then passed into the lower 'sub-cool' condenser where more gas can be transformed into liquid. Dual-core sub-cool condensers are very efficient.

Most condensers are made from aluminium as it is good at conducting heat and will not create a chemical reaction when it comes into contact with the refrigerant or lubricating oil.

Evaporators

An air conditioning system is a heat exchanger. It takes the heat energy from the passenger compartment and transfers it to the outside air. The component that is used to absorb the heat energy from inside the passenger compartment is the evaporator. The evaporator is constructed in a similar manner to a radiator, but instead of radiating heat as the design suggests, it creates a large surface area to absorb heat. (Heat energy will always travel from a hot substance to a cold substance). As the air in the passenger compartment is circulated through and around the evaporator core, heat energy is absorbed into the refrigerant. As with condensers, evaporators will normally be constructed with a serpentine or parallel design (see condensers).

The design and shape will allow for maximum air flow, while still creating a large surface area for heat absorption. The fins will also create a cold surface that will condense moisture from the air in the passenger compartment,

Diagnostics for Mobile Air Conditioning & Climate Control

helping to dehumidify it. Dirt, dust, smoke and pollen particles will stick to this damp surface helping to purify the air. As the water moisture and dirt build up on the fins, it will drip off and collect in the bottom of the ventilation unit where a tube will allow it to drain to the outside of that car.

Figure 3.12 An evaporator

The build-up of moisture and dirt on the fins of an evaporator provides a breeding ground for bacteria. This can lead to unpleasant smells and an environment that will make the passengers feel unwell (sick car syndrome). Some manufacturers will coat the outside surface of the evaporator with a thin film of silver because this has a natural anti-bacterial action which reduces the amount of cleaning required.

Receiver driers and suction accumulators

In order for the air conditioning system to operate efficiently, the compressor and condenser must be able to supply an excess of liquid refrigerant. This means that the refrigerant circuit needs a place to store the surplus refrigerant until required. Depending on the type of circuit used, this is done with either a receiver/drier or a suction accumulator.

In a thermal expansion valve TXV system, a receiver/drier is used which is mounted just after the condenser in the high pressure circuit. It is a cylindrical metal container with unions on the top which will allow it to be connected in series with the high pressure circuit. The inlet union vents directly into the top of the receiver/drier, but the outlet will have a pipe that will extend into the bottom of the receiver/drier known as a 'dipper tube' so that it draws on liquid refrigerant and not gas which may be at the top.

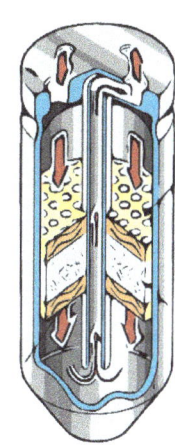

Figure 3.13 Receiver drier

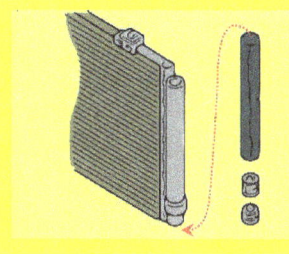

Some manufacturers have produced a cartridge type receiver/drier which slots into the side of the condenser housing. The inlet is fed directly from the condenser but the outlet will still contain a dipper tube to ensure that the system only draws on liquid refrigerant.

Diagnostics for Mobile Air Conditioning & Climate Control

In a fixed orifice tube FOT system, a suction accumulator is used which is mounted just after the evaporator in the low pressure circuit. It is a cylindrical metal container with unions on top which will allow it to be connected in series with the low pressure circuit. In a fixed orifice tube system, refrigerant is continuously flowing into the evaporator, and operation and pressure are regulated by cycling the compressor clutch on and off. Because of its constant operation there is the possibility that some refrigerant will remain in a liquid state as it passes into the low pressure circuit. If this liquid refrigerant was able to make it back to the compressor, it is possible that this would cause the compressor to hydro-lock. (Liquids are virtually incompressible and the pumping mechanism of the compressor would no longer be able to operate). The inlet of the suction accumulator has a pipe that extends into the bottom of the container allowing any liquid refrigerant to gather there. The outlet is vented into the low pressure return/suction pipe of the compressor from the top of the accumulator so that only gaseous refrigerant returns.

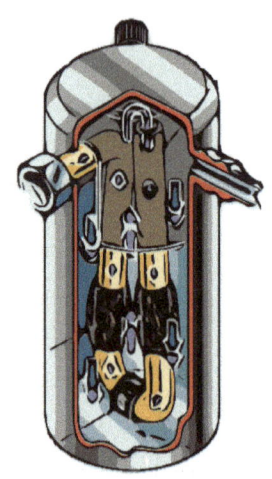

Figure 3.14 Suction accumulator

Both the receiver/drier and suction accumulator are also designed to trap dirt that that could contaminate and damage the air conditioning system. They will also both contain a silicon desiccant to help remove any water moisture from the circuit. Silicon is a naturally occurring substance, that when processed into a granular or beaded form, will have a strong attraction for water. The silicon desiccant is able to trap any water moisture in the receiver/drier or suction accumulator and help prevent it from circulating in the system where it may cause corrosion or blockages through the formation of ice.

An air conditioning system is sealed when in use and the only time it might be exposed to water moisture is during maintenance or repair. If the system has been evacuated and components disconnected, always ensure that any orifices are plugged to prevent exposure to air which may be carrying water moisture.

The silicon desiccant in a suction accumulator or receiver/drier can only absorb around 5 -10 ml of water moisture. Once it has become saturated, any further water will be free to circulate in the system and might cause damage. A good indication that water moisture is present in an air conditioning system can be seen when undertaking gauge diagnostics. If the desiccant in a receiver/drier or suction accumulator becomes saturated, it is not normally repairable and the whole component will require replacement.

In some older systems a receiver/drier or suction accumulator may be fitted with a small window known as a 'sight glass'. This sight glass can sometimes be used for basic diagnosis.

Diagnostics for Mobile Air Conditioning & Climate Control

Fusible plugs and pressure relief valves

If the temperature of the condenser or high pressure circuit rises too much there is the possibility that pipes or hoses may burst causing damage and refrigerant leakage. On some older systems a' fusible plug' or 'melt bolt' was used as a safety device. A fusible plug is a metal bolt, often screwed into the top of a receiver/drier that contains a material with a melting point of around 100 to 110°C. If system temperatures rose beyond this point, the fusible plug would melt and relieve pressures by venting refrigerant gas to atmosphere. Due to restrictions brought about because of the environmental damage caused by refrigerants, fusible plugs are no longer used. Instead high pressure relief valves are fitted to the compressor or sometimes to the receiver/drier.

Thermal expansion valves TXV

A thermal expansion valve TXV is able to regulate the flow of refrigerant into the evaporator due to system temperature and pressure.

There are three main types of thermal expansion valve:

Externally equalised thermal expansion valve - this type of valve has a regulating unit controlled from a diaphragm. A plunger is held against a seat in the closed position by a spring, meaning that no flow of refrigerant takes place. The plunger is attached to a diaphragm which can be actuated by the expansion of a liquid or gas contained in a **capillary tube**. The heat sensing capillary tube is mounted on the evaporator and the amount that the liquid or gas expands or contracts is dictated directly by the temperature of the evaporator.

As the evaporator cools, the pressure inside the capillary tube falls. This reduces the pressure on the diaphragm and plunger. The spring is able to act against the plunger reducing the flow of refrigerant into the evaporator.

As the evaporator warms up, the pressure inside the capillary tube rises. This increases the pressure on the diaphragm and plunger. The plunger is pushed against the spring force increasing the flow of refrigerant into the evaporator.

An externally equalised thermal expansion valve will also have a passageway or pipe connecting the lower chamber of the diaphragm to the low pressure circuit of the evaporator. The equalising tube will balance the reaction of the diaphragm due to pressures created in the evaporator, making this a very accurate method of refrigerant regulation.

Figure 3.15 A thermal expansion valve TXV

Internally equalised thermal expansion valve - this type of valve has a regulating unit controlled from a diaphragm. A plunger is held against a seat in the closed position by a spring, meaning that no flow of refrigerant takes place. The plunger is attached to a diaphragm which can be actuated by the expansion of a liquid or gas contained in a capillary tube. The heat sensing capillary tube is mounted on the evaporator and the amount that the liquid or gas expands or contracts is dictated directly by the temperature of the evaporator.

As the evaporator cools, the pressure inside the capillary tube falls. This reduces the pressure on the diaphragm and plunger. The spring is able to act against the plunger reducing the flow of refrigerant into the evaporator.

Diagnostics for Mobile Air Conditioning & Climate Control

As the evaporator warms up, the pressure inside the capillary tube rises. This increases the pressure on the diaphragm and plunger. The plunger is pushed against the spring force increasing the flow of refrigerant into the evaporator.

The lower diaphragm chamber of an internally equalised thermal expansion valve is joined to the high pressure circuit before the evaporator. This is a simpler design but not as accurate as an externally equalised valve because it doesn't regulate refrigerant flow depending on evaporator pressure.

Box type thermal expansion valve - this type of valve has the same pressure and temperature sensing functions of an externally equalised TXV but no external capillary tube. Instead the refrigerant entering the evaporator passes through a variable inlet in the box valve which is regulated by the temperature of the refrigerant passing through an outlet in the box valve from the evaporator.

Liquid refrigerant enters the evaporator through a small ball valve where its pressure falls and it boils, absorbing latent heat lowering the temperature of the evaporator. As the cold refrigerant leaves the evaporator, it passes back through the box valve and cools an internal thermal regulator which reacts against the inlet ball valve and reduces the flow.

As the flow of refrigerant into the evaporator falls its temperature increases slightly, and the exiting refrigerant passes back through the box valve and warms an internal thermal regulator. The regulator then reacts against the inlet ball valve, and increases the flow of refrigerant into the evaporator.

Capillary tube - a tube which has an internal diameter of hair-like thinness.

Helical – a spiral shape.

No matter which thermal expansion valve is used in an air conditioning system, it must be replaced with the same type as the original if required during maintenance and repair.

Fixed orifice tubes FOT

Another design that can be used for metering the flow of refrigerant into the evaporator is a fixed orifice tube FOT. It is located in the high pressure circuit after the condenser, normally close to the evaporator. The location of the fixed orifice tube can often be determined due to the size of a section in the high pressure pipe, or when in operation, at the point where the pipe goes from hot to cold. Refrigerant passes through a small mesh filter and then on to a tube with a calibrated size restriction. After passing through the fixed orifice tube, the refrigerant then enters the evaporator where the pressure falls and it boils, absorbing heat from the passenger compartment. With a FOT system, the only thing that regulates the quantity of refrigerant passing into the evaporator is system pressure. The

Diagnostics for Mobile Air Conditioning & Climate Control

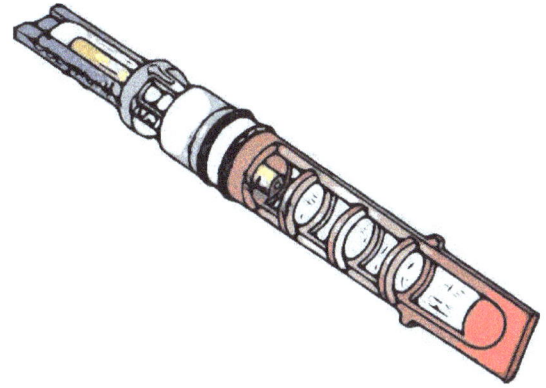

regulation of pressure is normally controlled by turning the compressor on and off. This is achieved by cycling the compressor clutch in and out depending on the pressure in the low side, sensed by a switch mounted in the evaporator. This type of system is very difficult to regulate accurately and under low loads, the evaporator may become exposed to excess liquid refrigerant. Because of this it is necessary to have an accumulator after the evaporator to ensure that no liquid refrigerant manages to get all the way back to the compressor.

Figure 3.16 A fixed orifice tube FOT

Some manufacturer's employ a variable displacement compressor in their design for use with fixed orifice tube systems to help regulate system pressures (see swash plate compressors).

Another design that can be used in this system type is a smart orifice valve which is able to alter the size of the orifice depending on temperature. In this type of regulator, an extra component is incorporated which can alter the size of the orifice tube. A plunger mechanism is held inside the orifice by a **helical** bimetallic spring which can react to system temperature and control the flow of refrigerant into the evaporator.

Variable orifice tubes can be fitted as an aftermarket upgrade to a standard FOT system, but special tools will be required for their replacement and manufacturer's recommendations should always be followed.

Hoses, pipes and service ports

For a description of hoses, pipes and service ports see Chapter 2.

Diagnostics for Mobile Air Conditioning & Climate Control

Control devices

Depending on whether the system is air conditioning (manually controlled) or climate control (automatically controlled) a number of options are available to manage the function and operation of the refrigerant cycle.

Figure 3.17 The controls of an air conditioning/climate control system

In an air conditioning or climate control system, the driver will normally have the following options:

Temperature control - in an air conditioning system this will usually be in the form of a dial or slide which can vary the electrical resistance in the temperature sensing electrical circuit of the air conditioner. By changing the electrical circuit resistance, the operation of the compressor can be controlled to regulate the evaporator temperature or move flaps in the ventilation circuit to control the flow of air through the evaporator and heater matrix. With climate control the desired temperature is set by the driver using dashboard controls and the cabin temperature is then monitored by a sensor in the passenger compartment. The sensor sends information to an ECU which then operates actuators to regulate the compressor and evaporator flow, ensuring that the set temperature is maintained.

Air distribution - in an air conditioning system this will usually be in the form of a dial or slide which can operate flaps within the HVAC system to direct air flow to various outlets or vents. The options will normally include windscreen, face and feet but depending in vehicle design other auxiliary vents may be included. The control will operate the flaps via cables, vacuum pipes and servos or small electric motors. With climate control, the flaps will operate in a similar manner to those found in air conditioning, but will normally only use electric servo motors to provide the positioning of the flaps. An ECU is able to manage the positioning of the flaps depending on how the driver has set the controls on the dashboard. During rapid warm up or cool down of the passenger compartment the ECU may override the drivers setting by initially directing warm air down (warm air rises) and cool air upwards (cold air falls). This way cabin temperature can be achieved in the quickest possible time, before the flaps return to the driver settings.

Ventilation fan speed - in an air conditioning system this will usually be in the form of a dial or slide which controls the power to an electric motor, driving a fan unit. Many air conditioning systems will use a **rheostat,** which sends power through a set of dropping resistors connected in series with the electric motor to reduce the voltage available.

Diagnostics for Mobile Air Conditioning & Climate Control

Rheostats will often give the option of 3, 4 or 5 speeds. An example of a three speed fan operation is described in the next section.

- **Off** - the dial is in the off position, causing an open circuit and no current flows to the motor.

- **One** - the electric motor feed is directed through a resistor in the rheostat which converts a large amount of the energy into heat. For example: if the resistor uses up 6 volts, then only 6 volts will be available at the motor giving slow speed.

- **Two**- the electric motor feed is directed through a resistor in the rheostat which converts some of the energy into heat. For example: if the resistor uses up 3 volts, then 9 volts will be available at the motor giving medium speed.

- **Three** - the electric motor feed bypasses the rheostat completely, meaning that 12 volts is available at the motor giving full speed.

A good indication that the rheostat speed control unit has failed in the ventilation fan unit is that only full speed is available when the control is operated.

Rheostat - an electrical component designed to control current flow by varying resistance.

Transistor- an electronic component which acts like a switch with no moving parts.

With climate control it is often possible to set ventilation fan speed stages in a similar manner to air conditioning, but many systems will often have a greater selection of settings or a dial/slider which gives variable control. In these systems, speed control is adjusted via the ECU using duty cycle. Duty cycle is an electricity regulating method that uses **transistors** to rapidly switch a component on and off, which unlike using resistors can operate with very little loss of power. This rapid switching does not allow the fan motor to fully speed up or slow down and depending on the duty cycle's 'on' time power to the motor can be controlled between an upper and lower limit giving various speeds. Using an ECU and duty cycle allows the climate control system to manage the air flow to the passenger compartment, for example:

- During rapid warm up or cool down, a high fan speed can be used to get the cabin to the desired temperature quickly.
- At varying road speeds ventilation can be regulated so that the forced air effect caused by the moving car can be compensated for. (i.e. at high road speeds the fan can be slowed down to allow for the air being rammed into the passenger compartment by momentum and when the vehicle is stationary the fan speed can be increased to maintain desired ventilation).

Diagnostics for Mobile Air Conditioning & Climate Control

Recirculation

Recirculation is another form of ventilation control. It is often operated by a different button or dial than the rest of the dashboard controls. When operated a flap acts to block off the outside air vents that allow ventilation in by the lower edge of the windscreen scuttle. This can be used to prevent unpleasant smells or fumes from entering the passenger compartment. In a climate control system, the recirculation flap can be automatically activated to assist with rapid warm up, cool down or demisting. An ECU can also use information provided by an air purity sensor to control its operation in polluted situations.

Climate control sensors

For a climate control system to automatically manage the heating, ventilation and air conditioning (HVAC), it needs to gather information from various sensors. A description of some of these sensors is shown in table 3.4.

Table 3.4 Climate control sensors

Sensor	Purpose and operation
Ambient air temperature sensor 	The ambient air temperature sensor is mounted in a position where it is able to take a measurement of the outside air temperature. It is a small **thermistor** which may be a **NTC** or **PTC** type. As outside temperature rises or falls, the resistance of the sensor changes and information is relayed to the ECU as a varying voltage. The ECU will then compare the information provided by this sensor with values of cabin air temperature and adjust the loading on the climate control system to compensate. This ensures that the most efficient operation of the heating, ventilation and air conditioning is achieved.
Cabin air temperature sensor  	The cabin air temperature sensor is mounted in a position where it is able to take a measurement of the passenger compartment temperature. It is often located inside a small pipe, called an aspirator tube that can be connected to the **plenum chamber** of the ventilation fan. This aspirator tube is then able to create a low pressure which helps draw cabin air across the sensor. The sensor is a small thermistor which may be a NTC or PTC type. As inside temperature rises or falls, the resistance of the sensor changes and information is relayed to the ECU as a varying voltage. The ECU will then compare the information provided by this sensor with values of ambient air temperature and the desired temperature set by the driver. It will then adjust the loading on the climate control system to compensate. This ensures that the most efficient operation of the heating, ventilation and air conditioning is achieved while maintaining a constant temperature inside the cabin.
Coolant temperature sensor 	Climate control systems will blend warm air that has passed through the heater matrix with purified/dehumidified air that has passed through the evaporator. In order for the ECU to accurately control the temperature which has been set by the driver, it needs to know the temperature of the coolant in the heater matrix. This measurement can be obtained from the engine management coolant temperature sensor, which is often a form of NTC thermistor, and then shared with the climate control ECU via network communication.

Diagnostics for Mobile Air Conditioning & Climate Control

Table 3.4 Climate control sensors

Sensor	Purpose and operation
Evaporator temperature sensor 	The evaporator temperature sensor is mounted in a position where it is able to take a measurement of the evaporator core temperature. It is a small thermistor which may be a NTC or PTC type. As evaporator temperature rises or falls, the resistance of the sensor changes and information is relayed to the ECU as a varying voltage. The ECU will then compare the information provided by this sensor with values of cabin air temperature and adjust the loading on the climate control system to compensate. The refrigerant cycle is carefully controlled from information provided by this temperature sensor in order to keep the evaporator core at just above 0°C. If this is not done, it would be possible for the condensed water moisture from the air to freeze on the outside of the evaporator core, preventing air flow and reducing overall efficiency of the system
Condenser temperature sensor 	To ensure correct operation of the refrigerant cycle, the condenser must efficiently dissipate heat to the surrounding air. This means that the condenser temperature must be kept within certain limits. At times of high system load and when ambient temperature is also high, extra air flow may be required in order for the condenser to radiate heat. The sensor will be mounted in a position where it is able to measure the temperature of the condenser core and operate a cooling fan if it becomes too hot.
Sun load sensor 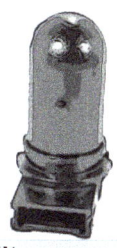	The sun load sensor is a small photoelectric cell. It is normally mounted on the dashboard of the car behind the windscreen. In direct sunlight you will feel hotter than the actual ambient cabin temperature due to the radiation produced by the sun's rays. It is the job of the sun load sensor to pick up this added solar radiation and adjust the loading on the climate control system to compensate, regardless of passenger compartment temperature.
Air quality sensor 	The air quality sensor is mounted in the main ventilation air intake system. It uses a chemical reaction to indicate the presence of carbon monoxide or nitrogen dioxide which changes the resistance of the sensor. The resistance change is used by the ECU to detect pollution and regulate the position of the recirculation flap controlling the exchange of incoming air to the passenger compartment.
Engine speed sensor 	Information from the engine management systems engine speed sensor is often shared with the climate control systems ECU via network communication. The engine speed is normally registered by the crankshaft position sensor which may be inductive or hall effect. The information from this sensor can be processed so that the compressor clutch of the climate control system does not cut in immediately after start-up, but instead waits until the engine has been running for a short period of time to ensure that excessive loads are not placed on the engine. This provides a more stable start-up/warm-up period, reduces fuel consumption and emissions and can increase engine performance during acceleration.

Diagnostics for Mobile Air Conditioning & Climate Control

Table 3.4 Climate control sensors

Sensor	Purpose and operation
Vehicle speed sensor 	Information from the cars dynamic control system vehicle speed sensor is often shared with the climate control ECU via network communication. The vehicle speed is normally registered by the transmission speed sensor or a wheel speed sensor which may be inductive, MRE or hall effect. The information from this sensor can be processed so that the speed of the ventilation fan can be regulated. When the car is moving fast, more air is forced into the passenger compartment and the ventilation fan slows down. When the car is moving slowly or is stopped, less air enters the passenger compartment so the ventilation fan is speeded up.

Thermistor - a temperature sensitive resistor (thermal resistor).

NTC - negative temperature coefficient. When used with thermistors, as temperature rises, resistance falls.

PTC - positive temperature coefficient. When used with thermistors, as temperature rises, resistance rises.

Plenum chamber - an air storage chamber.

It is worthwhile remembering that the temperature of the air coming out of the dash vents may be considerably different from that set by the driver on the climate control display. This will be most apparent during warm-up or cool down when the climate control system is trying to either add or remove heat from the passenger compartment.

Diagnostics

Tooling

No matter what diagnostic task you are performing on an air conditioning or climate control system, you will need to use some form of tooling.

Always use the correct tools and equipment.

The following table shows a suggested list of diagnostic tooling that could be used when testing and evaluating light vehicle heating, ventilation and air conditioning systems. Due to the nature of complex system faults, you will experience different requirements during your diagnostic routines and so you will need to adapt the list shown for your particular situation.

Diagnostics for Mobile Air Conditioning & Climate Control

Table 3.5 Tools and equipment used for the diagnosis and repair of HVAC systems

Tool	Possible use
Recovery management station (RMS) 	To safely evacuate an air conditioning system of refrigerant without escape to atmosphere. Hold a vacuum to help determine if the system has a leak, and safely refill after repairs have been completed.
Manifold gauge set 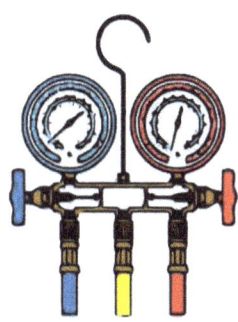	To connect to an operating refrigeration circuit and test system pressures.
Non-contact laser thermometers 	A non-contact thermometer can help determine the correct function and operation of an air conditioning or climate control system. It can be used to measure the temperature of air coming from the vents and also the temperature of the refrigerant circuit components.
Oxygen free nitrogen (OFN) 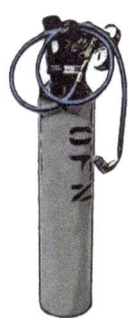	To pressurise a refrigerant circuit with a non-environmentally hazardous gas when checking for leaks.

Diagnostics for Mobile Air Conditioning & Climate Control

Table 3.5 Tools and equipment used for the diagnosis and repair of HVAC systems

Tool	Possible use
Refrigerant flushing equipment 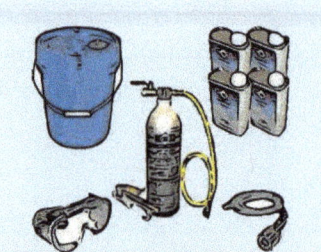	To clean a condenser that has been contaminated by a damaged compressor.
Dye injection kit and UV light 	To add fluorescent dye to an air conditioning system which can help locate the position of a leak with the aid of an ultraviolet light.
Oil injection equipment 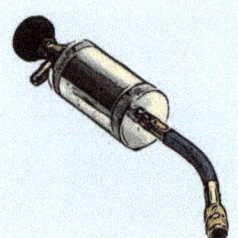	To add PAG or ester oil to the refrigerant circuit following the replacement of a condenser.
Refrigerant identifiers 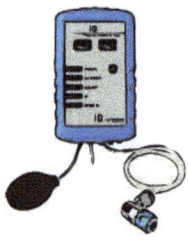	To help determine the type of refrigerant contained in a system before starting work and ensure that it doesn't contain hydrocarbons.
Test lamp/logic probe 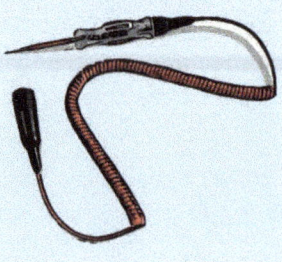	To test the existence of system voltage at a ventilation fan motor (Test lamps should be used with extreme caution on electronic systems, as the current draw created can severely damage components).

Diagnostics for Mobile Air Conditioning & Climate Control

Table 3.5 Tools and equipment used for the diagnosis and repair of HVAC systems

Tool	Possible use
Power probe 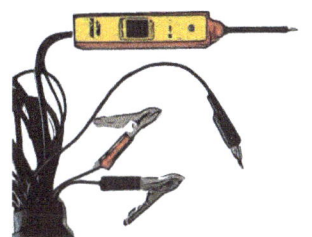	To power the electromagnetic clutch of a compressor and check its operation.
Multimeter 	To test the voltage or current draw of a climate control electrical circuit. It can also be used to check the resistance of system sensors.
Code reader/scan tool 	To retrieve diagnostic trouble codes (DTC) related to the climate control system. To clear trouble codes, reset the malfunction indicator lamp (if applicable), and evaluate the effectiveness of repairs.
Oscilloscope 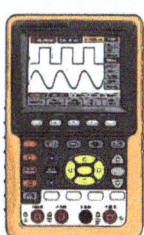	To test the signal produced by a multiplex network system which is communicating with the climate control.

Diagnostics for Mobile Air Conditioning & Climate Control

Electrical fault finding

Electronic and electrical safety procedures

Working with any electrical system has its hazards and you must take safety seriously. When you are working with light vehicle electrical and electronic systems, the main hazard is the possible risk of electric shock. Although most systems operate with low voltages of around 12V, an accidental electrical discharge caused by incorrect circuit connection can be enough to cause severe burns. Where possible, isolate electrical systems before repairing or replacing components.

If working on hybrid vehicles, take care not to disturb the high voltage system. The high voltage system can normally be identified by its reinforced insulation and shielding, which is often coloured bright orange. These systems carry voltages that can cause severe injury or death.

Always use the correct tools and equipment. Damage to components, tools or personal injury could occur if the wrong tool is used or misused. Check tools and equipment before each use.

If you are using electrical measuring equipment, you should check that it is accurate and calibrated before you take any readings.

If you need to replace any electrical or electronic components, always check that the quality meets the original equipment manufacturer (OEM) specifications. (If the vehicle is under warranty, inferior parts or deliberate modification might make the warranty invalid. Also, if parts of an inferior quality are fitted, it might affect vehicle performance and safety). You should only carry out the replacement of electrical components if the parts comply with the legal requirements for road use and environmental protection.

Electrical control principles of climate control systems

Table 3.6 describes some electrical principles and components involved with the operation of climate control systems.

Table 3.6 The operation of electrical and electronic systems

Electrical/Electronic system component	Purpose
ECU	The electronic control unit is designed to monitor and control the operation of light vehicle electrical systems. It processes the information received and operates actuators that control refrigeration, heating and ventilation systems for comfort and convenience.
Sensors	The sensors monitor various HVAC loads, components and ambient conditions against set parameters. As the driver makes demands on the climate control system, dynamic sensing creates signals in the form of resistance changes (that provide a varying voltage) which are relayed to the ECU for processing.
Actuators	The actuators are used to control the heating, ventilation and air conditioning systems operation. When operated by the ECU, motors, solenoids, valves, transformers, etc. help to control the action of the comfort and convenience systems.

Diagnostics for Mobile Air Conditioning & Climate Control

Table 3.6 The operation of electrical and electronic systems

Electrical/Electronic system component	Purpose
Digital principles 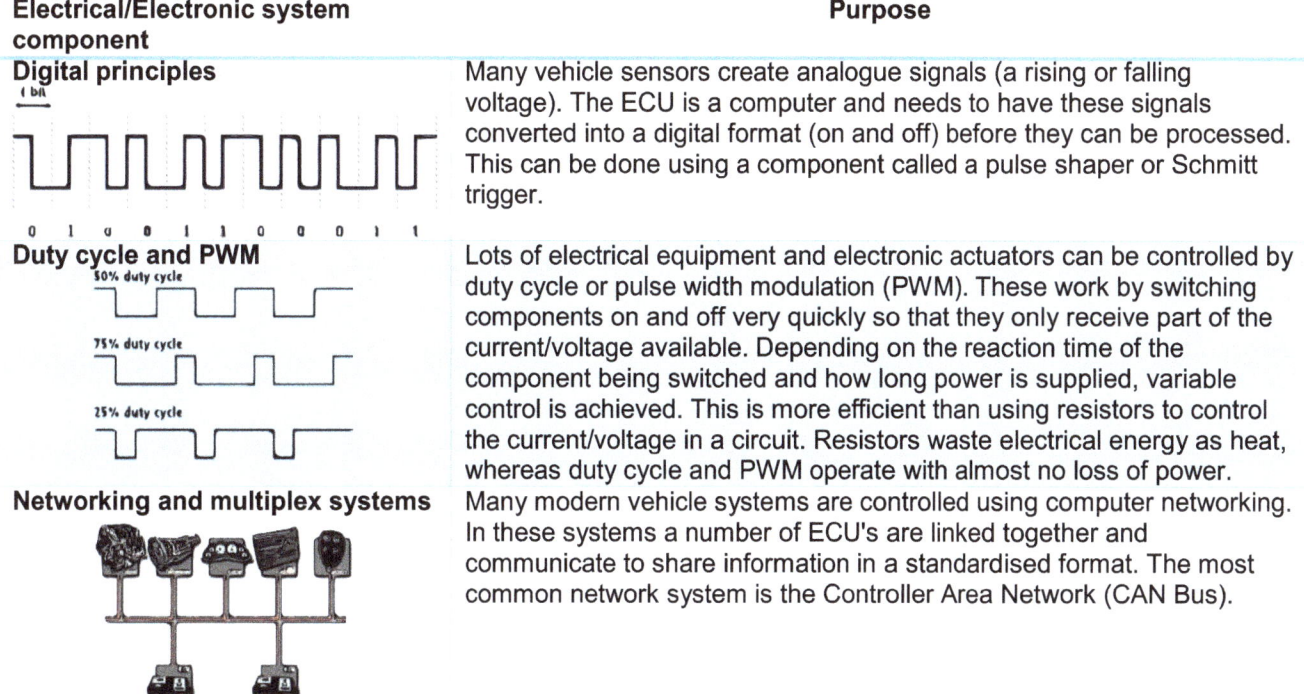	Many vehicle sensors create analogue signals (a rising or falling voltage). The ECU is a computer and needs to have these signals converted into a digital format (on and off) before they can be processed. This can be done using a component called a pulse shaper or Schmitt trigger.
Duty cycle and PWM	Lots of electrical equipment and electronic actuators can be controlled by duty cycle or pulse width modulation (PWM). These work by switching components on and off very quickly so that they only receive part of the current/voltage available. Depending on the reaction time of the component being switched and how long power is supplied, variable control is achieved. This is more efficient than using resistors to control the current/voltage in a circuit. Resistors waste electrical energy as heat, whereas duty cycle and PWM operate with almost no loss of power.
Networking and multiplex systems	Many modern vehicle systems are controlled using computer networking. In these systems a number of ECU's are linked together and communicate to share information in a standardised format. The most common network system is the Controller Area Network (CAN Bus).

Electrical principles related to light vehicle electrical circuits

You need to have a basic understanding of the principles of electricity before you can diagnose problems in an electrical circuit. This is crucial for anyone involved in the diagnosis and repair of automotive air conditioning or climate control electrical systems. This knowledge will allow you to make the right judgments and arrive at a successful conclusion, leading to a first time fix. Some basic electrical principles can be found in Chapter 2.

In cars, electrical energy is created by a chemical reaction (in a battery for example) or by the disruption of magnetic fields near electrical conductors (in a generator for example). You can measure how the electrical energy is created, moved and used using the electrical units shown in table 3.7.

Table 3.7 Electrical units

Volts	Voltage is electrical pressure. Voltage is the potential force in any part of an electrical circuit. Two main types of voltage occur in electrical circuits: Electromotive force (EMF) is potential pressure, and is usually considered to be the open circuit voltage when all electrical consumers are switched off and no current is flowing. It should be higher than electrical system voltage when current is flowing. Potential difference (Pd) is a circuit voltage measurement when components are switched on and current is able to flow. It is a measurement of voltage drop compared to the EMF at different positions within a circuit.

Diagnostics for Mobile Air Conditioning & Climate Control

Table 3.7 Electrical units

Amps	Amps are the units used to measure the amount of electricity in any part of an electrical circuit. Amps is measured when electricity is allowed to flow in an electrical circuit – this is known as current. There are two main types of electrical current: Direct current (DC) is electricity that flows in one direction only. Alternating current (AC) is electricity that moves backwards and forwards in an electric circuit. Amperage is the same wherever you measure it in the circuit (at the beginning, in the middle or at the end).
Ohms	Ohms are the units used to measure the resistance to electrical flow. Resistance has a direct effect on the operation of any electrical circuit. As resistance rises in a circuit, current and voltage fall, which can restrict the operation of electrical components. In some electrical circuits, resistance can be used as a method of control for electrical components, but in most circumstances a high resistance is undesirable.
Watts	Watts are the units used to measure electrical power made or consumed. Power is defined as the rate at which work is done. When referring to electrical components, the higher the wattage, the more powerful the component will be and the more electrical energy it will use.

How to use electrical diagnostic tooling

Electricity is invisible and as a result you will need to use specialist electrical diagnostic tooling to enable you to see what is happening in a circuit. Some examples of electrical diagnostic tooling and how to use them are described in the next section:

Test lamps

One of the simplest diagnostic tools you can use is a test lamp. Whether this is a professionally built tool or a bulb and a couple of pieces of wire that you have put together yourself, this tool can be very effective. Its purpose is to check to see if the circuit has power.

To use a test lamp on a low voltage system:

Figure 3.18 Test lamp

- **Step 1**: Connect one end of the test lamp to a good earth, such as the vehicle chassis or ground (the negative terminal of the battery is better because this is the end of all electrical circuits on a car).
- **Step 2**: Connect the other end of the test lamp to the part of circuit that needs to be checked.
- **Step 3**: If power exists in the circuit the test lamp will illuminate.

Diagnostics for Mobile Air Conditioning & Climate Control

Many modern test lamps have a point at the end of the probe to enable you to pierce wiring installation. Take care when using this probe as it is very easy to stab your finger. Also, if you pierce insulation, you are opening up the wiring to the effects of **oxidisation** from the air. If the wiring is left open, this oxidisation can lead to a high resistance and electrical problems in the future. It is best, where possible, to **back-probe** an electrical plug so as not to damage the wiring.

If you have to pierce the insulation of the wiring, you should cover it with insulation tape or heat shrink. Wiping a small amount of silicon into the pierced insulation hole is not advisable as it can cause corrosion in the wiring and high resistance.

Oxidisation – the effect of oxygen on metal, which can cause corrosion.

Back-probe – a method of making a test connection at the back of an electrical socket or plug.

Every time you connect another consumer to an electrical supply wire, more electrical current will be drawn from that supply wire until eventually it can take no more. A test lamp contains a bulb and this is a consumer. A standard test lamp has a low resistance, usually around 6 ohms. This means that when testing a low voltage vehicle electrical circuit, 2 to 3 amps of electrical current may be drawn. If a test lamp is used on an electronic circuit, severe damage can be caused as this high amperage moves through the components.

Always take care when using test lamps to diagnose electrical faults on vehicles. They should only be used when it is safe to do so. It is far safer to use an LED test light, if you are likely to be testing near electronic circuits.

Power probe

A power probe is an advanced form of test light, with additional features and capabilities. Power probes are usually fitted with light emitting diodes, LEDs that are able to illuminate in different colours when connected to either a powered circuit (LED glows red) or an earth circuit (LED glows green).

Checking polarity

After a simple connection to the vehicle's low voltage battery, you are able to see quite easily whether a circuit is positive or negative, without having to change **polarity** from one battery terminal to another. The power probe normally comes with two crocodile clips (red and black) – connect these to the appropriate positive and negative battery terminals.

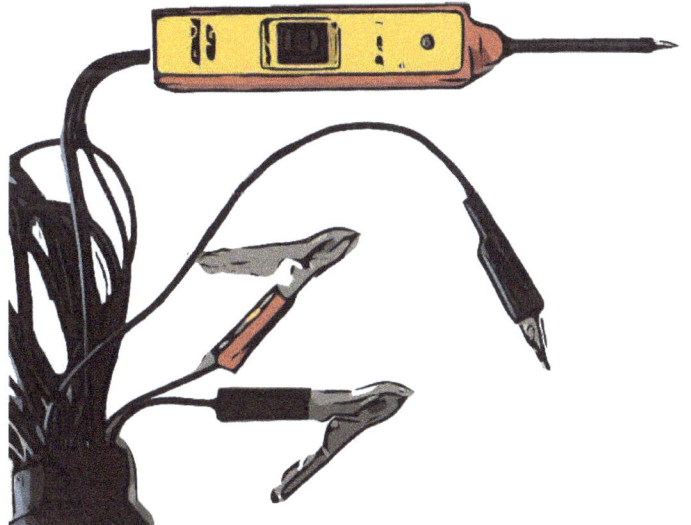

Figure 3.19 A power probe

Diagnostics for Mobile Air Conditioning & Climate Control

Polarity – a term used to describe electrical connection to a circuit. It represents the positive and negative connections.

Auxiliary – something that functions in a supporting capacity.

To check that a correct connection has been made, quickly touch the tip of the power probe to each battery terminal in turn:

- The LED should illuminate red when touched to the positive terminal.
- The LED should illuminate green when touched to the negative terminal.

Checking continuity using a power probe

Not only can a power probe check for electrical feed and earth, you can also use it to check for continuity (a continuous or unbroken conductor). You can check continuity on wires or components that have been disconnected from the vehicle's electrical system.

The power probe has an **auxiliary** ground wire – connect this to one end of the conductor, wire or component. Connect the tip of the power probe to the other end. If continuity exists, the LED on the power probe will illuminate.

Always remember to turn off power first before disconnecting a wire or component on the vehicle's electrical circuit.

Conducting functional tests using a power probe

How to:

You can also use the power probe to undertake functional tests of electrical components.

Step 1
- It is recommended that you disconnect the component from the vehicle's electrical system when conducting this test.

Step 2
- Connect the auxiliary ground to one terminal of the component and the tip of the power probe should be connected to the other.

Step 3
- Check that the LED illuminates to show that the component has continuity.

Step 4
- Keeping an eye on the LED, quickly rock the power switch and immediately release.

Diagnostics for Mobile Air Conditioning & Climate Control

Step 5 • If the LED indicator changed momentarily from green to red, you may proceed with the test.

Step 6 • By rocking the power switch forwards and holding it down, electrical potential will be supplied to the component and you can check its operation.

Step 7 • If during the initial rocking of the power switch the LED turned off, this normally indicates that the current being drawn by the component is too high for the power probe and the internal circuit breaker has tripped. This may require a manual reset and you will need to check the manufacturer's instructions.

Safety: Power probes are only designed to test components which draw relatively small amounts of current. Never use them to test starter motors, vehicle drive motors, etc.

Multimeters

The multimeter is a piece of electrical test equipment designed to measure a number of different units within an electrical circuit. There are two types of multimeter: analogue and digital.

Figure 3.20 A multimeter

Analogue multimeters

Analogue multimeters use a needle that moves across a graduated scale to record electrical readings within a circuit. The old-fashioned name for this type of unit was an 'AVO meter', which stood for amps, volts and ohms.

The problem with analogue meters is that they are only as good as the operator. The graduated scale can be difficult to read and so inaccurate readings could be obtained. Depending on the range of the scale provided by the manufacturer, a needle that lies somewhere between two units could be reading any fraction available. Analogue multimeters also have an upper range limit. If the needle flicks all the way to the end of this scale, it is known as full-scale deflection (FSD).

Figure 3.21 An analogue multimeter

Diagnostics for Mobile Air Conditioning & Climate Control

Digital multimeters

Digital multimeters shows digits (numbers) on a liquid crystal display (LCD) screen. These numbers are clearly displayed and are easy to read accurately.

> It is quite normal for the last digit on the far right of the screen to continuously change. This is a feature common to most digital multimeters. In a lot of cases, as really high accuracy is not required, this figure can be ignored.

Digital multimeter types

Two types of digital multimeter are common: manually operated and autoranging. With a manual multimeter, the operator selects the unit and the scale to be measured, normally by turning a dial on the front of the multimeter.

Using a manual multimeter

When you are using a manual multimeter, if you do not know the scale to be used, always follow this procedure:

- For testing volts and amps, first select the highest scale on the dial, then rotate the dial slowly down through the scales until you obtain an accurate reading.
- For testing ohms, first select the lowest scale on the dial and then rotate the dial slowly up until you obtain an accurate reading.

Figure 3.22 A manual multimeter

Using an autoranging multimeter

With an autoranging multimeter, the operator selects the unit but the scale of that unit is automatically selected by the multimeter. When using an autoranging multimeter, you must be careful that your reading is accurate by taking note of the scale of the unit being displayed.
For example, if voltage is measured, the scale might be in:

- Millivolts
- Volts
- Kilovolts
- Megavolts

Figure 3.23 An autoranging multimeter

Diagnostics for Mobile Air Conditioning & Climate Control

Using a digital multimeter

You can measure a number of electrical units on a digital multimeter, including volts, amps and ohms, but other measurements can also be taken.

Extra facilities on a digital multimeter may include: temperature, frequency, diode testing, transistor tests and audible continuity testing.

The electrical units of volts and amps are often broken down into two further areas: DC ⎓ and AC∼.

- The DC scale is normally shown on the meter as a straight line with a number of dots underneath it ⎓. This symbol is designed to prevent confusion. If just a single line was used, it might be mistaken for a minus sign and if two lines were used it might be mistaken for an equals sign.
- The AC scale is normally shown on the meter as a wavy line ∼.
- The ohms scale on a multimeter is normally represented by the Greek letter omega (Ω) because if the letter 'O' was used, it might be confused with zero.

Using a multimeter to check voltage

You can use a multimeter as a voltmeter to measure the pressure difference in an electric circuit between where you place the black probe and where you place the red probe.

How to:

Step 1
- Connect the probes to the correct sockets on the front of the multimeter.
- Connect the black lead and test probe to the common socket.
- Connect the red probe and test lead to the voltage socket.

Step 2
- Most low voltage systems that you will measure on a light vehicle will use direct current DC, so select the scale with the straight and the dotted lines.

Step 3
- Connect the voltmeter in parallel.

Step 4
- Connect the tip of the black lead to a good source of earth, such as the battery terminal, metal bodywork or engine.

Step 5
- Use the tip of the red lead to probe the electrical circuit being tested.

Using a multimeter to check for electrical resistance

You can use a multimeter as an ohmmeter to measure resistance. When checking for electrical resistance, always make sure that the power is switched off first and disconnect the component to be tested from the circuit.

How to:

Diagnostics for Mobile Air Conditioning & Climate Control

Step 1
- Connect the probes to the correct sockets on the front of the multimeter.
- Connect the black lead and test probe to the common socket.
- Connect the red probe and test lead to the socket marked with the omega symbol (Ω).

Step 2
- Before you take any measurements, you need to calibrate the ohmmeter to check that it is accurate. Turn the selector dial to the lowest ohms setting and join the tips of the two probes together.

Step 3
- When the leads are connected, the readout should show zero or very nearly zero. (If any figures are shown on the screen you will need to add or subtract them from your final results).

Step 4
- When the leads are disconnected you should see OL (meaning off limits) or the number 1, which is used to represent the letter 'I' (meaning infinity).

Step 5
- Now connect the ohmmeter in parallel across the components so that you can measure the resistance.

You can also use the ohmmeter to check for continuity.

To check a piece of wire for continuity - Place the red and black probes at each end of the wire. The screen should display a very low resistance reading.

To check a switch for correct operation - Connect the red and black probes across the terminals and operate the switch.

In the off position, the display should read OL (off limits) or infinity.

In the on position, the reading on the display should be very close to zero.

Using an ohmmeter to check for high resistance in an electrical circuit can be misleading. For example: a 12 volt lighting circuit drawing 10 amps with a bad connection causing a 0.1 Ω resistance will reduce the overall circuit voltage by 1 volt (ohms law 0.1 Ω x 10 A = 1 V), but the same 0.1 Ω resistance in a starter motor circuit drawing 100 amps will reduce the overall circuit voltage by 10 volts (ohms law 0.1 Ω x 100 A = 10 V).
This low resistance in two different styles of circuit may have little or no effect in the lighting circuit but cause complete failure of the starter circuit.
It is far better to use a volt drop test when checking for a high resistance as this will give a clearer indication of why a circuit is not working correctly.

Using a multimeter to measure electrical current

When measuring the electrical current in a circuit use the amps setting on the multimeter, so that it is used as ammeter. Take care when using an ammeter because, if it is connected incorrectly, the multimeter can be damaged.

How to:

Diagnostics for Mobile Air Conditioning & Climate Control

Step 1
- Connect the probes to the correct sockets on the front of the multimeter.
- Connect the black lead and test probe to the common socket.
- Connect the red probe and test lead to the socket used for measuring amps. (This socket is normally separate from the one used to measure volts or ohms).

Step 2
- Turn the selector dial to amps measurement.

Step 3
- You need to break into the circuit being tested, being careful to avoid short circuits.

Step 4
- Connect the ammeter in series, turn on the circuit and measure the current.

A good place to connect an ammeter is at the fuse box – remove the fuse completely and replace it with the ammeter.

Never connect an ammeter in parallel (across a circuit). A good ammeter has a very low internal resistance, so if the ammeter is connected in parallel a short circuit is created, causing excessive current flow and the ammeter will be damaged. Also remember that, depending on the quality of your ammeter, the amount of current that you can measure may be restricted to around 10 amps.

Other functions of a multimeter

Many multimeters are capable of other functional tests in addition to checking voltage, amperage and resistance. Some examples of extra functions are described in the next section.

Audible continuity testing

Some multimeters include an audible continuity tester. This means that you can test the continuity of an electrical component without having to look at the screen.

- Connect the test probes to the multimeter: black to the common or ground socket and red to the ohms socket.
- Turn the dial to the audible continuity test setting.
- To calibrate the meter and check correct operation, touch the probes together. You should hear an audible tone.
- As with ohms testing, you must switch off circuit power and remove the component being checked from the circuit.
- Now connect the red and black probes to the terminals of the conductor. If continuity exists, you will hear the audible tone.

Diagnostics for Mobile Air Conditioning & Climate Control

Diode testing

Most multimeters include a diode test facility. A diode is a one-way valve for electricity. Conduct the test in a similar manner to the continuity test.

- Connect the test probes: black to the common or ground socket and red to the ohms socket.
- Turn the dial to the diode testing setting.
- To calibrate the meter and check correct operation, touch the probes together. The display should show an ohms reading of zero.
- As with ohms testing, you must switch off circuit power and remove the diode from the circuit. You may need to unsolder the diode to remove it.
- With the diode removed, connect the probes to the terminals. If the diode is operating correctly, the display should show a low ohms reading.
- When the polarity of the probes is swapped over the display should show an off limits or infinity reading.
- If it shows zero in both directions, the diode has become short circuited.
- If it shows off limits or infinity in both directions, the diode has become open circuited.

Figure 3.24 A diode symbol

Frequency testing

Some multimeters have a **frequency** test facility. Frequency is a measurement of how quickly a circuit switches. The reading is normally measured in **Hertz** (Hz). 1Hz is equal to one complete cycle of operation (on and off for example) occurring in one second.

- Connect the test probe leads to the appropriate sockets on the multimeter.
- Turn the dial to the frequency setting.
- Test the component while the circuit is operating.

> **Frequency** – how often something happens.
>
> **Hertz** – a measurement of frequency.
>
> **Transistor** – an electronic component which can operate as a switch or amplifier (with no moving parts).

Temperature measurement

Some multimeters have a temperature measurement facility. This normally requires an additional probe to be connected. The temperature probe usually has its own socket for connection. Once you have turned the dial to the appropriate setting, you can measure temperature by placing the end of the probe where the measurement is to be taken. (Temperature measurement can be useful for diagnosing air conditioning faults).

Diagnostics for Mobile Air Conditioning & Climate Control

Transistor testing

Some multimeters have a **transistor** testing facility. This facility is rarely used by automotive technicians. Transistors are small electronic switches with no moving parts. They are normally soldered to an electrical circuit board and have three connections: collector, emitter, and base.

There are two types of transistor in common use: positive negative positive (PNP) and negative positive negative (NPN). If the multimeter has a transistor test facility, a six connector socket will be available, marked PNP or NPN. The transistor must be unsoldered from its circuit and connected to one of these diagnostic sockets. The transistor can now be tested by following the multimeter manufacturer's instructions.

Inductive amps measurement

Using an ammeter to check electric current is intrusive and the circuit must be broken. Also, incorrect connection may cause damage to your ammeter. For these reasons, an alternative method of testing for amperage has been developed; some multimeters come with an inductive amps clamp, or this can be purchased separately as an additional unit.

If the inductive clamp is an additional add-on to a standard multimeter, the wires will be connected to the voltage sockets and the voltage scale selected (but will be used to represent amps displayed on the screen).

The amps clamp uses **electromagnetic interference (EMI)** to measure current flow within a circuit. It does not require connection in series but is simply clamped around the wire to be tested. When the circuit is switched on and current flows, you can read the amperage measurement from the display. (Make sure that you read the manufacturers operating instructions to know how to connect and read the current clamp).

This is not always as accurate as connecting an ammeter in series, but is quicker and should not cause damage if connected incorrectly. It is also able to take much higher amperage readings than a standard multimeter.

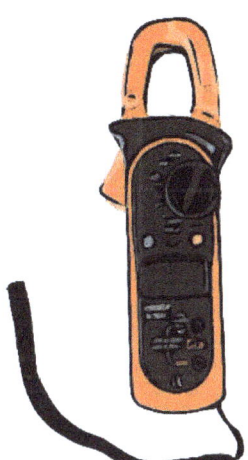

Figure 3.25 Inductive ammeter

Electromagnetic interference (EMI) – a disturbance that affects an electrical circuit due to either electromagnetic conduction or electromagnetic radiation emitted from an external source. It is also called radio frequency interference (RFI).

There is normally a plus or minus sign on the amps clamp to show which way round it should be connected to an electric circuit.

Oscilloscopes

An oscilloscope is a piece of electrical test equipment designed to act like a voltmeter or an ammeter. A multimeters measurement readout can't change fast enough to deal with modern electronic systems on motor vehicles – the numbers on the screen can't keep up. The answer to this is to use an oscilloscope.

Unlike a voltmeter, oscilloscopes not only show volts or amps but also time. Instead of a digital readout, the results are shown as a graph of volts or amps against time on a screen (as shown in Figure 3.27).

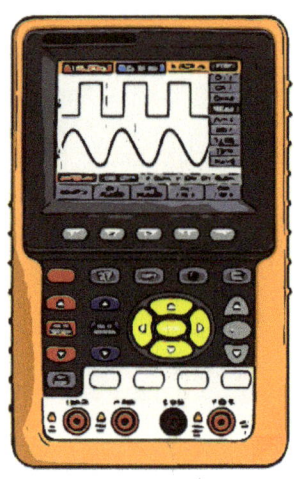

Figure 3.26 Handheld oscilloscope

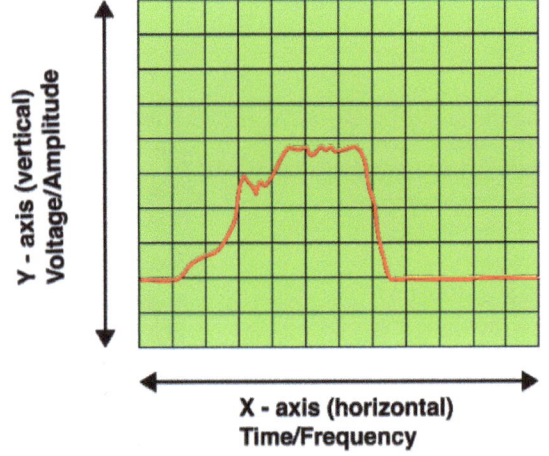

The graph normally shows voltage or amperage at the side of the screen (on the y-axis) – this axis is often called **amplitude**. Use the scale setting switch in a similar way to the dial on a manual multimeter to choose the amount of volts or amps that are shown on the screen.

Figure 3.27 An oscilloscope screen

An easy way to remember which axis is which on a graph is to say 'X is across' (a cross).

Diagnostics for Mobile Air Conditioning & Climate Control

- The graph normally shows time across the bottom of the screen (on the x-axis). This axis is often called **frequency**. Use the timescale switch in a similar way to the dial that is used to choose the amount of volts on a multimeter.

Amplitude – the height of a waveform, measured in volts or amps.

Frequency – the time scale of a waveform (how often something happens).

If you don't know what voltage or timescale to use on an oscilloscope, find out in the same way as you would with a multimeter. Start with the highest setting available and work downwards until you can see an image on the screen.

Lots of people are put off using oscilloscopes by the large box containing many wires and connectors. They feel that it will be complicated and time consuming to set up, so they don't bother.

However, to use an oscilloscope for simple electrical testing, you only need two probes – a common and voltage wire – just like with a multimeter. To measure amperage, you may need an inductive clamp.

Most of the diagnostic sockets found on oscilloscopes are colour-coded, so after a quick check of the manufacturer's instructions, it should be fairly easy to know where to plug these probes in.

How to:

Note: The oscilloscope probes may come in different colours, but for the sake of simplicity we will call them red and black here.

Step 1
- Connect the tip of the black lead to a good source of earth, such as the battery terminal, metal bodywork or engine. This will then only leave you with the red wire to worry about.

Step 2
- Now connect the red probe to the circuit to be tested.

Step 3
- Adjust the scales until you see an image on the screen.

Step 4
- After some practice, you will become familiar with the patterns and waveforms created by different vehicle systems.

Diagnostics for Mobile Air Conditioning & Climate Control

Scan tools and fault code readers

Faults with many modern vehicle climate control electronic systems would be difficult to diagnose without the aid of a scan tool. The electronic processes that take place within electrical and electronic circuits mean that these systems are being controlled many thousands of times a second, and faults can occur so quickly that you could miss them.

Some manufacturers include on-board diagnostic (OBD) systems as part of their vehicle design. The computer that controls the vehicle's climate control system may have a self-diagnosis feature. This allows it to detect certain faults and store a code number. Because the electronic control unit (ECU) is monitoring functions, it is able to record intermittent faults and store them in a keep alive memory (KAM) for retrieval by a diagnostic trouble code (DTC) reader.

It is a common misunderstanding to think that plugging a fault code reader into the vehicle's OBD system will tell you what the fault is. It actually only points you in the direction of the fault. You must test the system and components to find the fault.

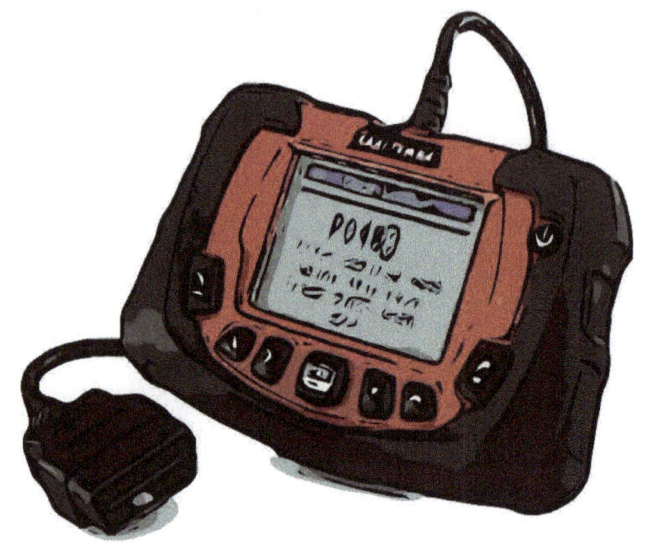

Figure 3.28 Scan tool

Common electrical faults

Diagnosing electrical faults can be confusing, as the symptoms can be very wide and varied. If you follow a simple approach, the diagnosis can be reduced to four main electrical faults:

- Open circuit
- High resistance (including bad earth)
- Short circuit
- Parasitic drain

Open circuit

In an open circuit, electricity cannot flow. This is normally because there is a physical break in the system. As a potential difference Pd will only occur in a circuit when current can flow, the fault can be diagnosed using a test method known as volts drop.
To diagnose an open circuit, you can use the multimeter as a voltmeter. Once set up correctly and connected to the appropriate circuit, measurements can be taken.
If the circuit is working properly, you should see full voltage all the way up to the consumer, at which point the electrical pressure should be used up. In an open circuit, the voltage will disappear. Using a voltmeter you can see at what point in the circuit this happens.
For example, Figure 3.29 shows using a voltmeter to check a low voltage open circuit in which the bulb does not light up. The voltmeter is connected at various points of the circuit to find out the voltage at these points. Where the voltage is different (between 12 volts and 0 volts), this shows the position of the open circuit. In this example, the open circuit is between points B and C.

Diagnostics for Mobile Air Conditioning & Climate Control

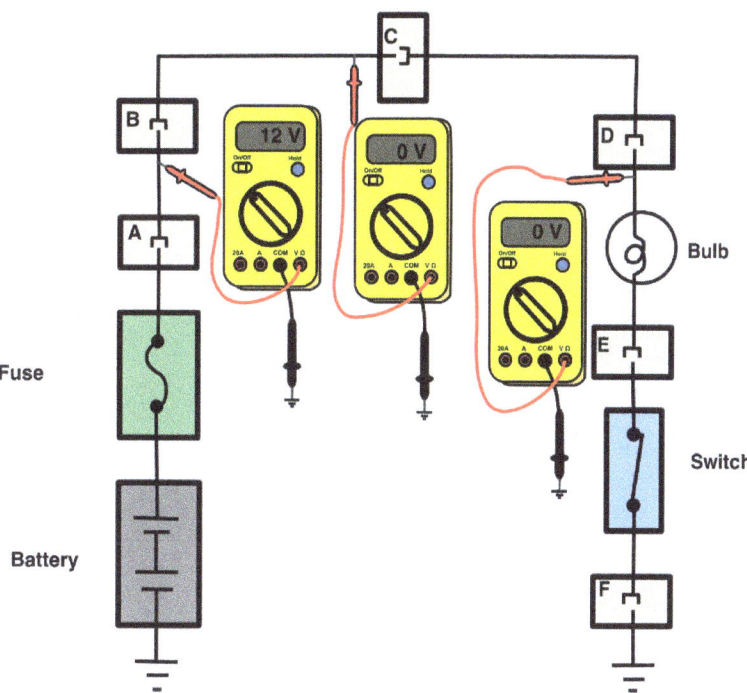

Figure 3.29 Testing for an open circuit

High resistance

In a high resistance circuit, the electricity slows down. This is normally because of a partial restriction in the system. Many high resistance faults are caused by poor, corroded or loose connections. A potential difference Pd will occur in a high resistance circuit, but the total Pd will be shared between the consumer and high resistance. This fault can also be diagnosed using the test method known as volts drop.
The symptoms of high resistance are that the component does not work properly (e.g. a bulb that glows dimly) because the circuit pressure (voltage) is shared with the resistance.
To diagnose this fault, you can use the multimeter as a voltmeter. Once set up correctly and connected to the appropriate circuit, measurements can be taken.
If the circuit is working properly, you should see full voltage all the way up to the consumer, at which point all the electrical pressure should be used up. In a high resistance circuit, the full voltage potential difference is not used up by the consumers. Using a voltmeter you can see at what point in the circuit this happens.
If there is a lower than expected voltage at the consumer, then the fault is in the first half of the circuit. If full voltage is present at the consumer, but some voltage still appears after the component, then the fault is in the second half of the circuit (bad earth).

A high resistance will make circuit current fall and this can sometimes be seen if an inductive amps clamp is connected to the circuit and the value compared with the fuse rating.

Diagnostics for Mobile Air Conditioning & Climate Control

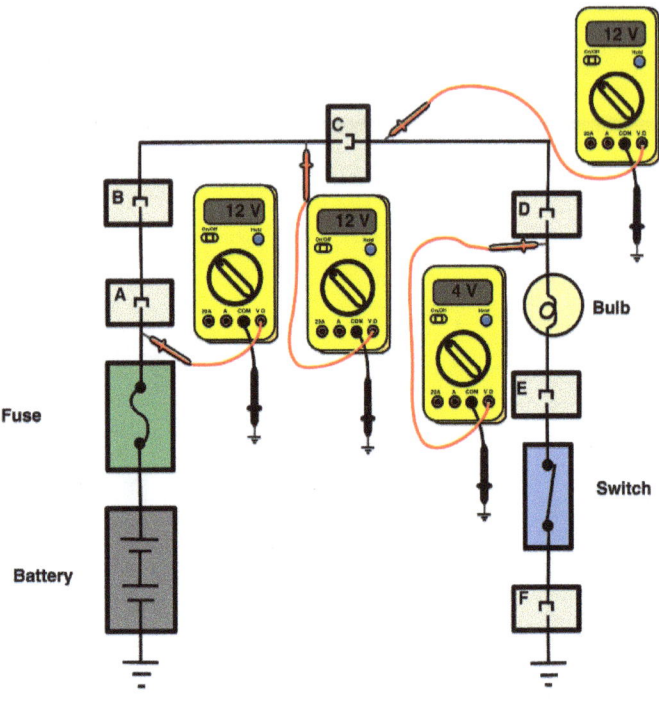

Figure 3.30 Testing for a high resistance

Bad earth

A bad earth is a high resistance after the consumer. If this exists the symptoms will be that the component won't work properly. Sometimes a bad earth can also cause the electrical energy to find an alternative path to the negative side of the battery.
To diagnose a bad earth use the same procedure as for high resistance.

Short circuit

Electricity is lazy, and will always take the path of least resistance. (Why travel the full length of the circuit when it can take a shortcut?)
In a short circuit, the electricity doesn't make it all the way to the end. Instead of going through the consumer, the electricity makes its way back to the battery early, and in the process converts its energy to heat.
The sudden discharge of current can cause a lot of damage, so the fuse that is used to protect the system should blow. If this happens the symptoms can make you think that the problem is an open circuit (which it is in a way, as the blown fuse has broken the circuit so that no current can flow).
In this situation, you can test the system with a voltmeter as explained in testing an open circuit, but once you have discovered the blown fuse, you should change your diagnostic routine to look for a short circuit. Any heat damage, including blown fuses, is a good indication that a short circuit might exist.

If a **dead short** to earth exists (e.g. the insulation of a wire has chafed against the metal bodywork of the vehicle), you can use a test lamp to help diagnose this fault (see Figure 3.31). It is important to use a test lamp containing a bulb and not an LED, as this could lead to system damage.
Once connected in place of the fuse, if the test lamp illuminates then the electricity is finding an alternative path back to the battery (short circuit). As the bulb is an electrical consumer, it uses up the electrical potential, and shouldn't damage the rest of the circuit. The circuit should then be disconnected systematically from the far end, working back towards the fuse box. When the bulb goes out, you have located the position of the dead short.

Diagnostics for Mobile Air Conditioning & Climate Control

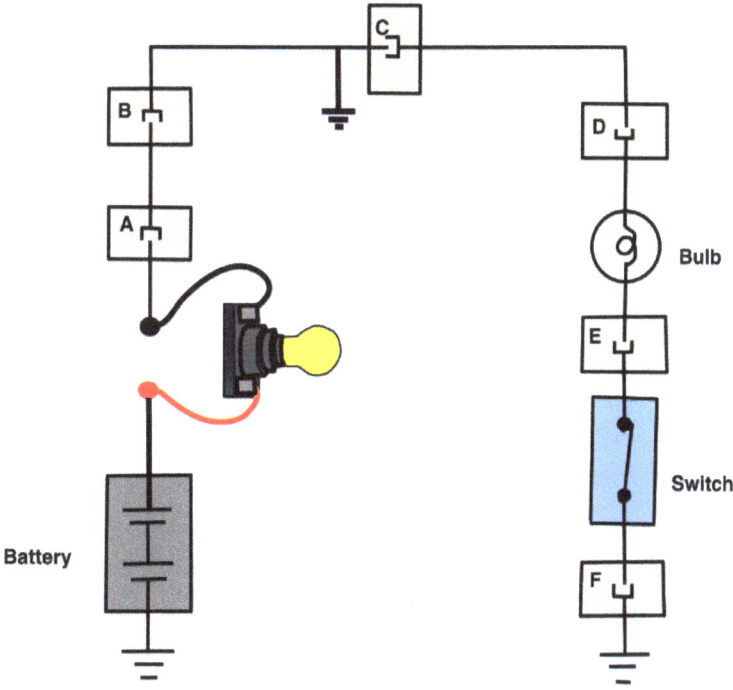

Figure 3.31 Testing for a short circuit

Dead short – an electrical short circuit that goes straight to earth without passing through a consumer.

A blown fuse may also be an indication of excessive current draw. With a test lamp connected as described for a short circuit test, an inductive amp clamp can be attached to the circuit (around the test lamp wire) and real time measurements can be taken by switching components on and off until the one with the high current draw is found.

Parasitic drain

A parasitic drain is similar to a short circuit – electricity will continue to flow even if the system is switched off, although this fault may not cause visible system damage. The symptom reported is normally that the battery goes flat if left for a period of time. To help diagnose this fault, you can use the multimeter or an inductive clamp meter as an ammeter. (This acts like a flow gauge to measure the amount of electric current moving in a circuit).

Diagnostics for Mobile Air Conditioning & Climate Control

Checking a parasitic drain

To check for parasitic drain, switch off all electric systems and connect the ammeter. The ammeter must be inserted into the circuit (connected in series) so that it isn't damaged. To do this you may need to disconnect one lead from the battery and use the ammeter to bridge the gap, so that current flows through it.

An inductive clamp can be placed around the battery wiring without disconnecting and inserting in series. This is much safer but is often less accurate than using an ammeter directly.

With everything switched off, there should be no current on the display of the meter. If any current (measured in amps) is shown, then a parasitic drain exists. To help find the parasitic drain, remove the fuses one at a time until **amps draw** falls to zero. This will help you locate the circuit containing the drain. Once you have identified the circuit, disconnect the components in that system until the current draw falls once again. You can now replace the faulty component.

The ammeter function of many multimeters will only provide readings of up to 10 amps before they are damaged. It is far safer where possible to use an inductive ammeter for testing.

Amps draw – the amount of current being used.

Table 3.8 Electrical testing

Electrical component	Example testing method(s)
Switches (12 volt circuit) 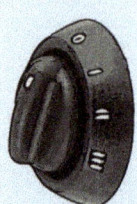	To test a switch removed from a circuit - isolate the power and disconnect the switch. Set up a multimeter to measure ohms and select the lowest setting. Calibrate the meter by touching the two probes together (the display should read zero). Connect the ohmmeter in parallel across the switch and turn on and off. In the on position the ohmmeter should display a very low reading close to zero. In the off position the ohmmeter should display infinity or off limits. To test a switch still connected to a circuit - set up a multimeter to measure volts and select the 20 volt DC setting. Connect the voltmeter in parallel across the switch and turn on and off. In the off position the volt meter should read 12 volts. In the on position the voltmeter should read 0 volts. (Any voltage shown on the display when the switch is closed is an unwanted volt drop and will reduce the operating performance of the circuit).

Diagnostics for Mobile Air Conditioning & Climate Control

Table 3.8 Electrical testing

Electrical component	Example testing method(s)
Sensors (variable resistance type) 	To test a variable resistance type sensor - set up an oscilloscope with a single channel. (The initial amplitude setting should be around battery voltage and the frequency set to quite a slow sweep). Connect the ground lead of the oscilloscope to a good earth point on the vehicle and the signal wire to the sensor being tested. (If you don't know which terminal on the sensor to connect to, try them all until a reading is obtained). Simulate the conditions that will make the sensor operate and check for a waveform on the display. After some practice you will become familiar with the type of waveform expected.
Relays (M4 type) 	To test a relay - it should be left connected to its circuit and access should be gained to the terminals on the base. Set up a multimeter to measure volts and select the 20 volts DC setting. Connect the ground probe of the voltmeter to a good vehicle earth point. Use the voltage probe to take measurements from the relay terminals. When the relay is operated, terminals 85, 30 and 87 should show battery voltage and terminal 86 should show 0 volts.
Connections 	To test an electrical connection - the circuit should be switched on and a multimeter set up to measure voltage. Choose the 20 volt DC setting and connect the voltmeter in parallel across the electrical connection. The display of the voltmeter should read as close to zero as possible. (Any voltage shown on the display is an unwanted volt drop and will reduce the operating performance of the circuit).
Fuses 	To test a fuse removed from a circuit - isolate the power and disconnect the fuse. Set up a multimeter to measure ohms and select the lowest setting. Calibrate the meter by touching the two probes together (the display should read zero). Connect the ohmmeter across the fuse. With a good fuse the ohmmeter should display a very low reading close to zero. If the fuse has blown the ohmmeter should display infinity or off limits. To test a fuse still connected to a circuit - set up a multimeter to measure volts and select the 20 volt DC setting. Connect the ground probe of the voltmeter to a good vehicle earth point. Use the voltage probe to take measurements at both ends of the fuse with the circuit switched on. With a good fuse, 12 volts will be displayed at both terminals. If the fuse has blown, one terminal will read 12 volts and the other will read 0 volts.
Motors (12 volt) 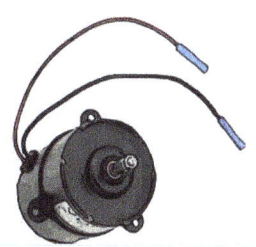	To test a motor still connected to a circuit - set up a multimeter to measure volts and select the 20 volt DC setting. Connect the voltmeter in parallel across the motor and switch the motor on. If operating, the volt drop shown on the display will indicate the amount of voltage being used by the motor (this should be as close as possible to 12 volts). If the motor does not operate, but 12 volts is displayed, this is an indication that the motor has failed.

Diagnostics for Mobile Air Conditioning & Climate Control

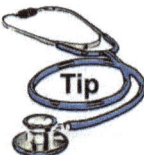

To help identify the amount of current a fuse is capable of withstanding before it blows, they are often colour coded. An example of the standardised colour codes are shown below:

Tan 5 amps

Brown 7.5 amps

Red 10 amps

Blue 15 amps

Yellow 20 amps

Clear 25 amps

Green 30 amps

DIN terminal numbers

To help technicians identify electrical terminals on vehicle circuits they are often given a standardised number. (For examples of these terminal numbers, see appendix in the back of this book).

Breakout boxes

To help with electrical diagnosis, some manufacturers supply breakout boxes. A breakout box is a piece of test equipment that enables the technician to measure electrical values at terminals or connectors without disconnecting the circuit. The breakout box fits in series with an electrical connector in the circuit to be tested and has a cable connected in parallel leading to a 'pin-out' box. The uninterrupted circuit can now be operated as normal and live measurements taken at the pin-outs using a multimeter or oscilloscope.

Refrigerant system gauge diagnosis

If you suspect a fault with the refrigerant circuit on an air conditioning or climate control system, it is often possible to carry out a diagnostic routine using the high and low pressure garages of a manifold set or recovery management station. With the pressure gauges connected (as described in Chapter 2) and the air conditioning operating the readings displayed on the dials can give an indication of the fault. Some examples of typical gauge readings and faults for a system using R134a are shown in table 3.9.

Diagnostics for Mobile Air Conditioning & Climate Control

Table 3.9 Pressure gauge diagnosis

Typical gauge reading	Possible fault/issue
Low pressure 21 - 35 psi. High pressure 199 - 228 psi	Normal
Low pressure 11 psi. High pressure 128 psi	Low charge/leak
Low pressure 35 psi. High pressure 327 psi	Overcharged
Low pressure 500mm/Hg - 21 psi. High pressure 99 - 213 psi (sudden flick between two different pressures)	Ice in system

Diagnostics for Mobile Air Conditioning & Climate Control

Table 3.9 Pressure gauge diagnosis

Typical gauge reading	Possible fault/issue
Low pressure 760mm/Hg. High pressure 85 psi	Blockage in system
Low pressure 57 - 85 psi. High pressure 100 - 142 psi (oscillating)	Compressor fault

Notes

Air should not be allowed to contaminate the refrigerant circuit and the system should only be filled once a vacuum has been created. Air is a non-condensable gas (meaning that it will not change state from gas to liquid) and as a result will reduce the effectiveness of the refrigeration cycle. Air in the system can sometimes be detected due to the noisier than expected operation of the air conditioning. If you suspect that there is air trapped in a refrigerant circuit you should conduct a full system evacuation and re-charge using a recovery management station (RMS). A recovery management station will often have an automatic air purge function.

Sight glass diagnostics

If the air conditioning system incorporates a sight glass, this can sometimes be used to perform a basic form of fault diagnosis. Some examples of refrigerant circuit operating conditions and the appearance of the sight glass are shown in table 3.10.

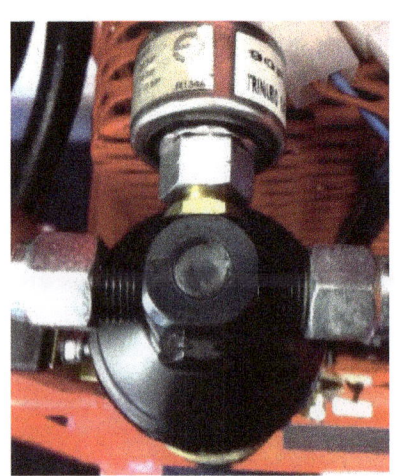

Figure 3.32 The sight glass of an air conditioning system

Diagnostics for Mobile Air Conditioning & Climate Control

Table 3.10 Sight glass diagnosis

Refrigerant circuit condition	Sight glass appearance and operation
System fully charged	Sight glass is clear
	Ventilation air is cold
	Compressor clutch is engaged
	Compressor inlet pipe is cool
	Compressor discharge pipe is warm
Very little or no refrigerant in system	Sight glass is clear
	Ventilation air is not cold
	Compressor inlet and discharge pipes are same temperature
System undercharged	Sight glass shows constant bubbles, foam, or oil streaks
Receiver drier is leaking desiccant	Sight glass becomes clouded with milky fluid

System pressure and temperature relationship

The pressures contained in an operating refrigeration circuit are a good indication of efficiency when compared to evaporator and ambient temperatures. Table 3.11 shows some approximate values for the relative pressures and temperatures with a system containing R134a which can be used as a reference when testing air conditioning and conducting maintenance.

Table 3.11 System pressure and temperature relationship

Low side readings		High side readings	
Low side gauge psi	Evaporator temperature °C	High side gauge psi	Ambient temperature °C
16 - 29 psi	0.5 - 10 °C	115 - 200 psi	21.1 - 26.6 °C
19 - 39 psi	0.5 - 15.5 °C	140 - 235 psi	26.6 - 32.2 °C
25 - 43 psi	4.4 - 18.3 °C	165 - 270 psi	32.2 - 37.7 °C
37 - 51 psi	8.8 - 18.3 °C	210 - 310 psi	37.7 - 43.3 °C

Checking for leaks using oxygen free nitrogen OFN

If you suspect that an air conditioning system is leaking, you should not attempt to fill the circuit with refrigerant as this would cause environmental pollution. Instead the system should be pressurised with oxygen free nitrogen OFN to help locate any leaks. Once the system has been pressurised a soap solution can be used to help find large leaks or fluorescent dye being expelled by the system can help find smaller leaks.

Diagnostics for Mobile Air Conditioning & Climate Control

How to:

Step 1
- Observing all health and safety requirements, connect the air conditioning recovery machine to the vehicle and fully evacuate the system.

Step 2
- Use a vacuum hold process to determine if the system has a leak.

Step 3
- Disconnect the recovery machine and connect the oxygen free nitrogen OFN to the high side service port.

Step 4
- Open the valve on the OFN gas cylinder and pressurise the system to 145 psi (10 Bar).

Step 5
- Use leak detection equipment to locate the source of the leak.

Checking for leaks using UV dye

One of the most common methods used for locating small leaks in an air conditioning system is by injecting a fluorescent dye into the refrigeration circuit and using an ultraviolet light to help highlight any dye expelled from the system. The procedure below describes how to locate a leak in a system that has dye in it.

How to:

Step 1
- Operate the air conditioning on full for a minimum of 10 minutes to enable the dye to fully circulate.

Step 2
- Switch off the engine and air conditioning.

Step 3
- Connect the ultraviolet light to a suitable power source and put on specialist yellow tinted UV glasses.

Step 4
- Use the ultraviolet light to help you locate any trace of dye found on components in the engine bay (compressor, condenser, TXV, FOT, hoses etc). Pay particular attention to joints and connections.

Step 5
- To check for any leaks inside the car you can often remove the pollen filter which may give you access to one side of the evaporator, but any further investigation will often involve stripping out the HVAC unit.

Diagnostics for Mobile Air Conditioning & Climate Control

Step 6
- If traces of dye are found, clean the dye away using an appropriate solvent, run the air conditioning on full for a further 10 minutes and recheck for dye to confirm the leak.

Step 7
- Fully evacuate the system following all health, safety and environmental procedures and repair the leak.

System performance test

Following any maintenance or repairs, and with the system fully charged, the air conditioning should be operated and tested for correct performance.

How to:

Step 1
- Observing all health and safety, connect the manifold gauges or recovery machine to the air conditioning systems high and low pressure service ports.

Step 2
- Lower the bonnet (but don't latch shut) being careful not to trap/damage the hoses or equipment.

Step 3
- Close all windows and doors and run engine at fast idle for approximately 5 minutes.

Step 4
- Connect a thermometer to one of the main dashboard vents, adjust air flow speed to low and place controls to maximum cooling.

Step 5
- After a 5 minute stabilisation period, check pressure gauges and readings compare with normal operating pressures.

Diagnostics for Mobile Air Conditioning & Climate Control

Step 6
- If the high side pressure reading becomes too high, an auxiliary fan can be placed in front of the condenser to simulate ram air flow and assist with heat transfer.

Step 7
- The thermometer in the dashboard vent should read 2 - 7°C at approximately 21 - 24°C ambient temperature. (note that higher ambient temperatures and high humidity will increase the outlet temperature).

Step 8
- If the system is operating close to normal temperature and pressure tolerances, stop the engine and disconnect gauges/recovery machine and refit service port dust caps.

Step 9
- If the system is not operating close to normal pressure and temperature tolerances, use the pressure gauge table 3.9 to help diagnose the issue.

If you are running an engine in a confined space or workshop, exhaust extraction must be used.

Faults and symptoms

With any diagnosis, it is important to listen to the symptoms described by the driver and work out a logical diagnostic routine before starting to work on the vehicle. Don't charge into a diagnosis, as this can lead to errors and misdiagnosis.

Always make sure that you fix the **fault** and not the **symptom**.

Fault – something that is responsible for an undesirable situation or event.

Symptom – a sign that indicates a certain situation, especially an undesirable situation.

Table 3.12 shows some examples of symptoms caused by faulty air conditioning and climate control components and operating conditions and their possible causes.

Table 3.12 Typical symptoms and possible causes

Symptoms	Possible cause
Evaporator temperature too low, with no control available.	TXV in stuck in an open position or orifice tube leaking.
System pressure rises too high and compressor clutch cuts out. Low pressure system pressure continuously falls.	TXV stuck in the closed position or orifice tube blocked.

Diagnostics for Mobile Air Conditioning & Climate Control

Table 3.12 Typical symptoms and possible causes

Symptoms	Possible cause
System fully charged but the compressor clutch won't cut in.	Pressure switch malfunction.
Poor air conditioning efficiency or compressor clutch not cutting in.	Refrigerant undercharged.
Compressor clutch not cutting in or constantly cycling on and off.	Refrigerant overcharged.
System pressure building up too high and then suddenly dropping to normal.	Water moisture in the system.
Inefficient and noisy operation of the refrigerant circuit.	Air in the system.
A temperature drop in the refrigeration circuit after the receiver drier.	Restriction in the receiver drier.
The suction accumulator gets warm during use.	Restriction in the suction accumulator.
Inefficient operation of the air conditioning system.	Blocked/damaged condenser fins.
Excessive noise from compressor when in operation.	Damaged compressor or insufficient lubrication oil.
Noise or damage to compressor drive belt.	Incorrect drive belt tension, loose/insecure mounting or misalignment of pulleys.
Reduced air flow and effectiveness of air conditioning.	Blocked pollen filter.
Different fan speeds unavailable.	Faulty fan motor rheostat/resistor unit
Loss of ventilation control.	Faulty servo motors on ventilation air distribution or recirculation flaps.
Incorrect temperature control.	Faulty passenger compartment temperature sensor/thermostat.
Inefficient operation of the air conditioning or reduced air flow through the ventilation system.	Blocked evaporator caused by ice build-up on the outer fins.
Water ingress.	Blocked or incorrectly routed evaporator drain hoses. Leaking heater matrix.
Vehicle odours.	Sick car syndrome caused by bacterial growth on the evaporator.

Routes to diagnosis

You need to develop a logical and systematic approach to your diagnostic routine in order to cut down on time, cost and frustration. Remember that correct diagnosis is important to ensure a first time fix.

The flow chart shown in Figure 3.33 gives an example of a generic routine that should be used when diagnosing a fault.

Diagnostics for Mobile Air Conditioning & Climate Control

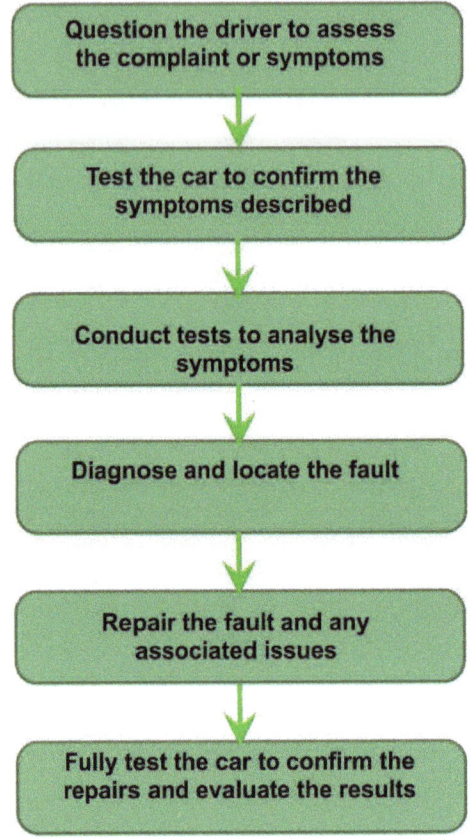

Figure 3.33 Generic diagnostic routine

The removal and replacement of faulty system components

Once a successful diagnosis has been conducted, you will need to repair the fault which may involve the replacement of system components. Before any repair or replacement is conducted you must safely evacuate the refrigerant from the air conditioning system following environmental precautions.

If you need to replace any of the refrigeration system components, always check that the quality of the parts meets the original equipment manufacturer (OEM) specifications. (If the vehicle is under warranty, inferior parts or deliberate modification might make the warranty invalid. Also, if parts of an inferior quality are fitted, this might affect vehicle performance and safety.) You should only carry out the replacement of air conditioning components if the parts comply with the legal and environmental requirements for road use.

Safe Environment

During the diagnosis and repair of mobile air conditioning systems (MAC), you may need to replace components that are contaminated with refrigerants. Under the Environmental Protection Act 1990 (EPA), you must dispose of these components in the correct manner. They should be safely stored in a clearly marked container until they are collected by a licensed recycling company. This company should give you a waste transfer note as the receipt of collection.

Diagnostics for Mobile Air Conditioning & Climate Control

Check your knowledge

1. If a climate control system is set to a high temperature, during the rapid warm up where will the majority air be distributed?
a Face
b Rear
c Feet
d Windscreen

2. If a fixed orifice tube is blocked, operating the A/C compressor will cause the low side pressure gauge to?
a Read an equal pressure to the high side pressure gauge.
b Rise constantly.
c Do nothing.
d Fall constantly.

3. Which unit of measurement do air quality sensors in climate control systems use to sense pollutant concentration in the air?
a Resistance
b Watts
c Volts
d Power

4. What is the correct procedure of measuring the resistance of a sensor with a multimeter?
a The circuit must be live.
b A current clamp must be used.
c The sensor must be disconnected from the circuit.
d The meter must be set to the 10 amp setting.

5. Which one of the following climate conditions places an extra cooling load on a climate control system?
a Sun
b Snow
c Wind
d Rain

Diagnostics for Mobile Air Conditioning & Climate Control

6. Which one of the following is not used on a manual air conditioning control system?
a Interior temperature sensor.
b Cables.
c Ventilation speed control.
d Electric motors.

7. The two values shown on the graph of an oscilloscope screen are:
a Voltage and power.
b Voltage and amperage.
c Voltage and time.
d Amperage and resistance.

8. What is the current rating for a yellow coloured fuse?
a 10 amp
b 15 amp
c 20 amp
d 25 amp

9. Which of the following is an actuator in a fully automatic climate control system?
a High pressure switch.
b Ambient temperature sensor.
c Sun load sensor.
d Blend flap motor.

10. What is the purpose of an ambient air temperature sensor?
a Measure outside air temperature.
b Measure evaporator temperature.
c Measure face vent temperature.
d Measure interior air temperature.

Answers: 1c, 2d, 3a, 4c, 5a, 6a, 7c, 8c, 9d, 10a

Diagnostics for Mobile Air Conditioning & Climate Control

Preparing for assessment

The information contained in this book can help you with theory or practical assessments used to identify your skills or competence when undertaking vehicle repairs or a recognised qualification. It is possible that some of the evidence you produce may contribute to more than one qualification. You should ensure that you make best use of all your evidence to maximise the opportunities for cross-referencing between units or qualifications.

You should choose the type of evidence that will be best suited to the type of assessment that you are undertaking (either theory or practical).

The types of evidence you could use are listed below:

- Direct observation by a qualified assessor
- Witness testimony
- Computer based
- Audio recording
- Video recording
- Photographic recording
- Professional discussion
- Oral questioning
- Personal statement
- Competence/Skills tests
- Written tests
- Multiple-choice tests
- Assignments/Projects

Before you attempt a written or multiple-choice test, make sure you have reviewed and revised any key terms that relate to the topics in that subject area. Make sure you read all questions carefully. Take time to digest the information so that you are confident about what the question is asking you. With multiple-choice tests, it is very important that you read all of the answers carefully, as it is common for two of the answers to be very similar, which may lead to confusion.

For practical assessments, it is important that you have had enough practice and that you feel that you are capable of passing. It is best to have a plan of action and work method that will help you.

Make sure that you have the correct technical information, in the way of vehicle data, and appropriate tools and equipment. It is also a good idea to check your work at regular intervals. This will help you to be sure that you are working correctly and to avoid problems developing as you work. When undertaking any practical assessment, always make sure that you are working safely throughout the test and following all environmental precautions.

Common Acronyms/Abbreviations

A - Amperes
A/C - Air Conditioning
A/F - Air Fuel Ratio
A/T - Automatic Transmission
AAC - Auxiliary Air Control Valve
AAT - Ambient Air Temperature
ABD - Automatic Brake Differential
ABS - Antilock Brake System
ABV - Air Bypass Valve
AC - Alternating Current
ACC - Automatic Climate Control
ACC - Air Conditioning Clutch
ACR - Air Conditioning Relay
ACR4 - Air Conditioning Refrigerant, Recovery, Recycling, Recharging
ACV - Air Control Valve
ADU - Analogue-Digital Unit
AEV - All Electric Vehicle
AFC - Air Flow Control
AFL - Advanced Front Lighting System
AFM - Air Flow Meter
AFR - Air Fuel Ratio
AFS - Air Flow Sensor
AGM - Absorbed Glass Matt
Ah - Amp Hours
AIR - Secondary Air Injection System
AIS - Automatic Idle Speed
ALC - Automatic Level Control
AM - Amplitude Modulation
API - American Petroleum Institute
APS - Atmospheric Pressure Sensor
ARC - Automatic Ride Control
ARS - Automatic Restraint System
ASARC - Air Suspension Automatic Ride Control
ATC - Automatic Temperature Control
ATDC - After Top Dead Centre
ATF - Automatic Transmission Fluid
ATS - Air Temperature Sensor
AVO - Amps Volts Ohms
AWD - All Wheel Drive
AWG - American Wire Gage
AYC - Active Yaw Control
B/MAP - Barometric/Manifold Absolute Pressure
BARO - Barometric Pressure
BCM - Body Control Module
BCM - Battery Control Module
BDC - Bottom Dead Centre
BEV - Battery Electric Vehicle
BHP - Brake Horsepower
BOB - Breakout Box
BP - Barometric Pressure
BPP - Brake Pedal Position Switch
BTDC - Before Top Dead Centre
BTS - Battery Temperature Sensor

Btu - British thermal unit
BUS N - Bus Negative
BUS P - Bus Positive
C - Celsius
CA - Cranking Amps
CAN - Controller Area Network
CANP - EVAP Canister Purge Solenoid
CAS - Crank Angle Sensor
CBW - Clutch by Wire
CC - Catalytic Converter
CC - Climate Control
CC - Cruise Control
CC - Cubic Centimetres
CCA - Cold Cranking Amps
CD - Compact Disc
CDI - Capacitor Discharge Ignition
CFC - Chlorofluorocarbons
CFI - Continuous Fuel Injection
CI - Compression Ignition
CKP - Crankshaft Position Sensor
CL - Closed Loop
CLC - Converter Lockup Clutch
CLV - Calculated Load Value
CMP - Camshaft Position Sensor
CNG - Compressed Natural Gas
CO - Carbon Monoxide
CO2 - Carbon Dioxide
COC - Conventional Oxidation Catalyst
COP - Coil on Plug Electronic Ignition
COSHH - Control of Substances Hazardous to Health
CP - Crankshaft Position Sensor
CP - Canister Purge (GM)
CPP - Clutch Pedal Position
CPU - Central Processing Unit
CRC - Cyclic Redundancy Check
CRD - Common Rail Diesel
CRS - Common Rail System
CTP - Closed Throttle Position
CTS - Coolant Temperature Sensor
CV - Constant Velocity
CVT - Continuously Variable Transmission
DBW - Drive by Wire
DC - Duty Cycle
DC - Direct Current
DCS - Dual Clutch System
DI - Distributor Ignition (System)
DI - Direct Ignition
DIS - Direct Ignition (Waste Spark)
DIS - Distributor-less Ignition System
DMF - Dual Mass Flywheel
DMM - Digital Multimeter
DLC - Data Link Connector (OBD)
DOHC - Dual Overhead Cam
DPF - Diesel Particulate Filter

Common Acronyms/Abbreviations

DRL - Daytime Running Lights
DTC - Diagnostic Trouble Code
DVD- Digitally Versatile Disc
EAIR - Electronic Secondary Air Injection
EBCM - Electronic Brake Control Module
EBP - Exhaust Back Pressure
EBD- Electronic Brake Force Distribution
ECC - Electronic Climate Control
ECM - Engine/Electronic Control Module
ECS - Emission Control System
ECT - Engine Coolant Temperature
ECU - Electronic Control Unit
EDC- Electronic Diesel Control
EECS - Evaporative Emission Control System
EEGR - Electronic EGR (Solenoid)
EEPROM - Electronically Erasable Programmable Read Only Memory
EFI - Electronic Fuel Injection
EFT - Engine Fuel Temperature
EGO - Exhaust Gas Oxygen Sensor
EGR - Exhaust Gas Recirculation
EGRT - Exhaust Gas Recirculation Temperature
EMF - Electromotive Force (voltage)
EMI - Electromagnetic Interference
EOBD - European On Board Diagnostics
EOP - Engine Oil Pressure
EOT - Engine Oil Temperature
EPA - Environmental Protection Act
EPB- Electronic Parking Brake
EPROM - Erasable Programmable Read Only Memory
EPS- Electronic Power Assisted Steering
ESP- Electronic Stability Programme
ESS - Engine Start-Stop
EVAP - Evaporative Emissions System
EVAP CP - Evaporative Canister Purge
FM- Frequency Modulation
FOT - Fixed Orifice Tube
FSD- Full Scale Deflection
FT - Fuel Trim
FWD - Front Wheel Drive
GDI - Gasoline Direct Injection
GND - Electrical Ground Connection
GPS- Global Positioning System
GWP - Global Warming Potential
H – Hydrogen
HASAWA- Health and Safety at Work Act
H2O - Water
HC - Hydrocarbons
HCA- Hot Cranking Amps
HDI- High Pressure Direct Injection
HEGO - Heated Exhaust Gas Oxygen Sensor
HFC- Hydrogen Fuel Cell

HFC- Hydro-fluoro Carbon
HFO- Hydro-fluoro Olefin
Hg – Mercury
HICE- Hydrogen Internal Combustion Engine
HID- High Intensity Discharge (lighting)
HO2S - Heated Oxygen Sensor
hp - Horsepower
HSE- Health and Safety Executive
HT - High Tension
HUD - Head up Display
HVAC - Heating Ventilation and Air Conditioning
Hz - Hertz
I/O - Input / Output
IA - Intake Air
IAC - Idle Air Control (motor or solenoid)
IAT - Intake Air Temperature
IC - Integrated Circuit
IC - Ignition Control
ICE- In Car Entertainment
ICE – Internal Combustion Engine
ICM - Ignition Control Module
IFS - Inertia Fuel Switch
IGBT- Insulated Gate Bipolar Transistor
IGN - Ignition
IGN ADV - Ignition Advance
IGN GND - Ignition Ground
IPR - Injector Pressure Regulator
ISC - Idle Speed Control
ISO - International Standard of Organisation
KAM - Keep Alive Memory
Kg/cm2 - Kilograms/ Cubic Centimetres
KHz - Kilohertz
Km - Kilometres
KPA - Kilopascal
KPI- Kingpin Inclination
KS - Knock Sensor
KWP - Keyword Protocol
l - Litres
LCD - Liquid Crystal Display
LED - Light Emitting Diode
LHD - Left Hand Drive
Li-ion- Lithium ion
LOOP - Engine Operating Loop Status
LOS - Limited Operating Strategy
LPG - Liquefied Petroleum Gas
LSD- Limited Slip Differential
LTFT - Long Term Fuel Trim
LWB - Long Wheel Base
M/T - Manual Transmission
MAC - Mobile Air Conditioning
MAF - Mass Air Flow Sensor
MAP - Manifold Absolute Pressure Sensor
MAT - Manifold Air Temperature

Common Acronyms/Abbreviations

MCM- Motor Control Module
MEF- Methane Equivalency Factor
MF- Maintenance Free
MFI - Multiport Fuel Injection
MIL - Malfunction Indicator Lamp
MPG - Miles per Gallon
MPH - Miles per Hour
mS or ms - Millisecond
mV or mv - Milivolt
N - Nitrogen
NCAPS - Non-Contact Angular Position Sensor
NCRPS - Non-Contact Rotary Position Sensor
NGV - Natural Gas Vehicles
Ni-MH- Nickel Metal Hydride
Nm - Newton Meters
NOx - Oxides of Nitrogen
NPN- Negative Positive Negative
NTC - Negative Temperature Coefficient
O2 - Oxygen
OBD I - On Board Diagnostics Version I
OBD II - On Board Diagnostics Version II
OC - Oxidation Catalytic Converter
OD - Overdrive
OD - Outside Diameter
ODP - Ozone Depletion Potential
OE - Original Equipment
OEM - Original Equipment Manufacturer
OFN - Oxygen Free Nitrogen
OHC - Overhead Cam Engine
OHV - Overhead Valve
OL - Open Loop
OS - Oxygen Sensor
P/N - Part Number
PAG - Polyalkylene Glycol
PAIR - Pulsed Secondary Air Injection
PATS - Passive Anti-Theft System
PCB - Printed Circuit Board
PCM - Powertrain Control Module
PCV - Positive Crankcase Ventilation
Pd-Potential Difference (volts)
PEF- Propane Equivalency Factor
PEM- Proton Exchange Membrane
PFI - Port Fuel Injection
PGM-FI - Programmed Gas Management Fuel Injection
PID - Parameter Identification Location
PKE - Passive Keyless Entry
PNP- Positive Negative Positive
POT - Potentiometer
PPE- Personal Protective Equipment
PPM - Parts Per Million
PPS - Accelerator Pedal Position Sensor
PROM - Programmable Read-Only Memory
PSI - Pounds per Square Inch

PTC - Positive Temperature Coefficient Resistor
PTO - Power Take Off (4WD Option)
PUWER- Provision and Use of Work Equipment Regulations
PWM - Pulse Width Modulation
RAM - Random Access Memory
RBS - Regenerative Braking system
RCM- Reserve Capacity Minutes
RDS - Radio Data System
REF - Reference
RFI - Radio Frequency Interference
RHD - Right Hand Drive
RIDDOR- Reporting of Injuries Diseases and Dangerous Occurrence Regulations
RKE - Remote Keyless Entry
RMS - Recovery Management Station
ROM - Read Only Memory
RON - Research Octane Number
RTV - Room Temperature Vulcanizing
RWD - Rear Wheel Drive
SAE –Society of Automotive Engineers (Viscosity Grade)
SAI- Swivel Axis Inclination
SCR- Selective Catalytic Regeneration
SCS - Sick Car Syndrome
SFI - Sequential Fuel Injection
SI- Spark Ignition
SIPS - Side Impact Protections System
SOC- State of Charge
SOHC - Single Overhead Cam
SPFI - Single Point Fuel Injection (throttle body)
SRI - Service Reminder Indicator
SRS - Supplementary Restraint System (air bag)
SRT - System Readiness Test
STFT - Short-Term Fuel Trim
SWB - Short Wheel Base
SWL- Safe Working Load
TAC - Throttle Actuator Control
TACH - Tachometer
TBI - Throttle Body Injection
TC - Turbocharger
TCC - Torque Converter Clutch
TCM - Transmission Control Module
TCS - Traction Control System
TD - Turbo Diesel
TDC - Top Dead Centre
TDI - Turbo Direct Injection
TOOT- Toe Out On Turns
TP - Throttle Position
TPM - Tyre Pressure Monitor
TPP - Throttle Position Potentiometer
TPS - Throttle Position Sensor
TSB - Technical Service Bulletin
TV - Throttle Valve

Common Acronyms/Abbreviations

TXV- Thermal Expansion Valve
UART - Universal Asynchronous Receiver-Transmitter
UJ- Universal Joint
USB - Universal Serial Bus
UV - Ultraviolet
V - Volts
VAC - Vacuum
VAF - Vane Airflow Meter
VDP- Variable Diameter Pulley
VDU- Visual Display Unit
VIN - Vehicle Identification Number
VPE- Vehicle Protection Equipment
VSS - Vehicle Speed Sensor
W/B - Wheelbase
WOT - Wide Open Throttle
WSS - Wheel Speed Sensor
YRS - Yaw Rate Sensor

Appendix
DIN Terminal Numbers

Ignition system

1	coil, distributor, low voltage
1a, 1b	distributor with two separate circuits
2	breaker points magneto ignition
4	coil, distributor, high voltage
4a, 4b	distributor with two separate circuits, high voltage
7	terminal on ballast resistor, to distributor
15	battery+ from ignition switch
15a	from ballast resistor to coil and starter motor

Preheat (Diesel engines)

15	preheat in
17	start
19	preheat (glow)

Starter

50	starter control

Battery

15	battery+ through ignition switch
30	from battery+ direct
30a	from 2nd battery and 12/24 V relay
31	return to battery- or direct to ground
31a	return to battery- 12/24 V relay
31b	return to battery- or ground through switch
31c	return to battery- 12/24 V relay

Electric motors

32	return
33	main terminal (swap of 32 and 33 is possible)
33a	limit
33b	field
33f	2. slow rpm
33g	3. slow rpm
33h	4. slow rpm
33L	rotation left
33R	rotation right

Turn indicators

49	flasher unit in
49a	flasher unit out, indicator switch in
49b	out 2. flasher circuit
49c	out 3. flasher circuit
C	1st flasher indicator light
C2	2nd flasher indicator light
C3	3rd flasher indicator light
L	indicator lights left
R	indicator lights right
L54	lights out, left
R54	lights out, right

AC generator (alternator)

51	DC at rectifiers
51e	as 51, with choke coil
59	AC out, rectifier in, light switch
59a	charge, rotor out
64	generator control light

Generator, voltage regulator

61	charge indicator (charge control light)
B+	battery +
B-	battery -
D+	dynamo +
D-	dynamo -
DF	dynamo field
DF1	dynamo field 1
DF2	dynamo field 2
U, V, W	AC three phase terminals

Lights

54	brake lights
55	fog light
56	spot light
56a	headlamp high beam and indicator light

Appendix
DIN Terminal Numbers

56b	low beam		83b	out position 2
56d	signal flash		**Relay**	
57	parking lights		85	relay coil -
57a	parking lights		86	relay coil +
57L	parking lights left		**Relay contacts**	
57R	parking lights right		87	common contact
58	licence plate lights, instrument panel		87a	normally closed contact
58d	panel light dimmer		87b	normally open contact
Window wiper/washer			88	common contact 2
53	wiper motor + in		88a	normally closed contact 2
53a	limit stop+		88b	normally open contact 2
53b	limit stop field		**Additional**	
53c	washer pump		52	signal from trailer
53e	stop field		54g	magnetic valves for trailer brakes
53i	wiper motor with permanent magnet, third brush for high speed		75	radio, cigarette lighter
			77	door valves control

Acoustic warning

71	beeper in
71a	beeper out, low
71b	beeper out, high
72	hazard lights switch
85c	hazard sound on

Switches

81	opener
81a	1 out
81b	2 out
82	lock in
82a	1st out
82b	2nd out
82z	1st in
82y	2nd in
83	multi position switch, in
83a	out position 1

Index

Actuators, 124

air quality sensor, 119

Alternating current (AC), 78, 125

ambient air temperature sensor, 118

Ambient temperatures, 13

Amplitude, 138

Amps, 81, 125

amps clamp, 136

Amps draw, 143

anthropogenic, 18

Armature, 93

Asphyxiation, 28

Atom, 78

Auxiliary, 127

Back-probe, 126

bad earth, 141

battery, 94

Boil off, 45

breakout box, 145

British Standards Institution (BSI), 62

British Thermal Units (BTU), 13

Brushes, 93

cabin air temperature sensor, 118

cabin filters, 59

Capillary tube, 114

Carry out or carry over, 45

CFC, 16

charging cylinder, 65

chlorofluorocarbons, 16

Circuit, 80

clutch, 107

Commutator, 93

Compressor, 5, 100

Condensable, 51

Condensation, 10

Condenser, 5, 108

Condenser temperature sensor, 118

Conduction, 9

Conductor, 78

Connector, 87

Consumer, 84

Continuity, 80, 128

COSHH control of substances hazardous to health, 22

Convection, 9

Convection currents, 57

coolant temperature sensor, 118

Cross contamination, 70

Current, 78

cycling, 108

Dead short, 142

Deflection gauges, 72

Deionised water, 95

Dichlorodifluoromethane, 16

diode, 135

Dipper pipe, 41

Direct current (DC), 78, 125

Duty cycle and PWM, 57, 124

Duty of Care, 21

earth return, 88

ECU, 124

Electrolyte, 95

Electromagnet, 91

Electromagnetic interference (EMI), 137

electromagnetic spectrum, 9

electromagnetism, 9

Electromotive force (EMF), 81, 125

electron flow, 79

Electrons, 78

engine speed sensor, 119

Environmental Protection Act 1990 (EPA), 21

Ester oil, 61

Evaporate, 10

Evaporator, 5, 110

evaporator temperature sensor, 118

Index

Expansion valve, 6
F gas, 20
filling weights, 43
first aid, 31
Fixed orifice tube FOT, 6, 7, 55, 114
Fluorescent dye, 67
fluoroelastomer, 30
flushing, 70
Frequency, 135, 138
Frostbite, 28
fuses, 94
fusible plug, 112
Generator, 80
Global warming, 17
global warming potential, 18
greenhouse effect, 18
Gross weight, 43
GWP, 18
Halide, 67
Harmonic tension, 72
Heat, 8
heat transfer, 8
Hertz, 135
high resistance, 140
Horse power, 59
Hoses, 6
Humidity, 9
hydro fluorocarbon, 17
Hydro-lock, 45
hygroscopic, 61
Insulator, 78
International Organization for Standardization (ISO), 62
joints, 54
Kelvin, 12
Kyoto agreement, 19, 20, 21
latent heat, 13
leak detection, 67
lubrication, 60

MAC, 20
Manifold, 64
Montreal Protocol, 19, 27
motor, 92
Mufflers, 53
Multimeters, 129
network, 118, 119, 124
Neutrons, 78
NTC, 118, 119
Nucleus, 78
OFN, 39, 67, 68, 96, 121, 149, 159
ohm, 81
Ohm's law, 81, 82
Ohms, 125
Oil injection, 66
open circuit, 140
oscilloscope, 137
Oxidisation, 126
oxygen free nitrogen, 39, 67, 149
ozone, 16
PAG oil, 60
Parallel circuit, 84
Parallel flow, 108
parasitic drain, 143
periodic table, 76
personal protective equipment, 3, 48
piston type compressor, 101
Plenum chamber, 120
Polarity, 127
pollen filter, 59
Pollen Sensor, 60
Potential difference (Pd), 81, 125
power probe, 127
PPE, 29
pressure, 13
Protons, 78
PTC, 118, 120
Purging, 70

Index

R12, 16, 25
R1234yf, 25
R134a, 24
R22, 25
R744, 15, 25
Radiation, 9
Receiver drier, 6, 111
Recirculation, 117
recovery management stations, 35
recovery position, 32
Refrigerant identifier, 34, 35, 69
refrigeration cycle, 50
Regulation (EC) No 842/2006, 20
Relative humidity, 10
Relay, 91
Resistance, 125
rheostat, 116, 117, 152
RMS, 35, 36, 63, 72, 120, 147, 159
Safety data sheet, 24
scotch yoke compressor, 103
scroll compressor, 104
Sensible heat, 12
Series circuit, 84
Serpentine, 108
Service connectors, 33
Short circuit, 95, 141
SI system, 62
sick car syndrome, 59, 60, 110
sight glass, 147
silicon desiccant, 112
Suction accumulator, 5, 111

sun load sensor, 119
superheat, 106
Swash plate compressor, 102
Symptom, 151
system recovery, 72
Tare weight, 43
Temperature, 11
Terminal, 87
test lamp, 125
Tetrafluoropropene, 19, 25, 35, 46
Thermal decomposition, 27
thermal expansion valve TXV, 113
thermal switch, 105
Thermistor, 119
thermometer, 68, 121
thermostat, 94
transfer note, 22
Transistor, 135
trinary switch, 107
TXV, 6, 14
ultraviolet, 16
vacuum pump, 65
vane type compressor, 101
vehicle protection equipment (VPE), 4, 49, 99
vehicle speed sensor, 119
ventilation, 56
Volt drop, 92
Voltage, 81
Volts, 125
Watts, 81, 125
weighing scales, 65

www.ingramcontent.com/pod-product-compliance
Lightning Source LLC
Chambersburg PA
CBHW061820290426
44110CB00027B/2928